International Series in Operations Research & Management Science

Volume 178

For further volumes:
http://www.springer.com/series/6161

Tsan-Ming Choi • Chun-Hung Chiu

Risk Analysis in Stochastic Supply Chains

A Mean-Risk Approach

Springer

Tsan-Ming Choi
Institute of Textiles and Clothing
The Hong Kong Polytechnic University
Hung Hom, Kowloon
Hong Kong

Chun-Hung Chiu
Department of Management Sciences
City University of Hong Kong
Kowloon Tong, Kowloon
Hong Kong

ISSN 0884-8289
ISBN 978-1-4614-3868-7 ISBN 978-1-4614-3869-4 (eBook)
DOI 10.1007/978-1-4614-3869-4
Springer New York Heidelberg Dordrecht London

Library of Congress Control Number: 2012940864

Printed on acid-free paper

Springer is part of Springer Science+Business Media (www.springer.com)

Preface

Risk analysis is one of the most crucial items on senior management's agenda. It is a critical part of modern supply chain management under a turbulent market environment. Over the past few years, there are more and more emerging works attempting to explore risk-related issues in stochastic supply chain management problems while the majority of them focus on conceptual framework or computational analysis. Pioneered by Nobel laureate Markowitz in the 1950s, the mean-risk formulation becomes a fundamental theory for risk management in finance. In recent years, there is growing popularity of applying this groundbreaking theory in analyzing stochastic supply chain management problems. Nowadays, there is no doubt that the mean-risk theory is a well-proven approach for conducting risk analysis in stochastic supply chain models. However, there is an absence of a comprehensive reference source that introduces the topic and provides the state-of-the-art findings on this influential approach for supply chain management. As a result, we coauthor this book and believe that this book will be a pioneering text focusing on this important topic.

This book comprises five chapters. Chapter 1 introduces the topic, offers a timely review of various related areas, and explains why the mean-risk approach is important for conducting supply chain risk analysis. Chapter 2 examines the single period inventory model with the mean-variance and mean-semi-deviation approaches. Extensive discussions on the efficient frontiers are also reported. Chapter 3 explores the infinite horizon multiperiod inventory model with a mean-variance approach. Chapter 4 investigates the supply chain coordination problem with a versatile target sales rebate contract and a risk-averse retailer possessing the mean-variance optimization objective. Chapter 5 concludes the book and discusses various promising future research directions and extensions. As a remark, every chapter can be taken as a self-contained article and the notation within each chapter is consistently employed.

In terms of the potential audience, we believe that this book is suitable for both researchers and practitioners in supply chain management. It can also be a good reference book for senior year undergraduate and postgraduate students. Since we target a rather broad pool of readers, we intentionally avoid the excessive usage of technical terms and try to use simple "layman" terms as much as possible. We believe that readers with the undergraduate level knowledge of calculus, probability, and statistics will be able to understand most of the technical aspects of the analysis.

Before closing, we would like to take this opportunity to show our gratitude to Fred Hillier, Neil Levine, and Matthew Amboy for their kind support and help along the course of carrying out this book project. Parts of this book are based on our prior published papers in Elsevier's journals, and we hence sincerely acknowledge Elsevier for granting us the authorship rights to reuse materials from our articles in this book. We are also indebted to our families, colleagues, friends, and students, who have been supporting us during the development of this book. This book is partially supported by the funding of the Research Grants Council of Hong Kong under grant number PolyU 5420/10H. Last but not least, the first author would like to thank his mentor Professor Duan Li for inspiring him to work on the mean-variance analysis of supply chain models a full decade ago. He also dedicates this book to celebrating Professor Li's 60th birthday in the coming July.

Hong Kong

Tsan-Ming Choi
Chun-Hung Chiu

Contents

About the Authors

Tsan-Ming Choi is an Associate Professor in Fashion Business at the Hong Kong Polytechnic University. His current research interests mainly focus on supply chain management and applied optimization. He is now an area editor/associate editor/editorial board member of over ten academic journals, which include *Decision Support Systems*, *European Management Journal*, *IEEE Transactions on Systems, Man and Cybernetics: Part A*, *Information Sciences*, and *Journal of Fashion Marketing and Management*. He has authored over 120 technical papers and published extensively in leading business and engineering journals such as *IEEE Transactions on Automatic Control, Production and Operations Management, Automatica, European Journal of Operational Research, Decision Support Systems,* and various other *IEEE Transactions*. His papers have also appeared in well-established fashion and textiles journals such as *Journal of Fashion Marketing and Management*, *Journal of the Textile Institute*, and *Textile Research Journal*. He is currently serving as an executive committee member and treasurer of Production and Operations Management Society (Hong Kong Chapter), and IEEE Systems, Man and Cybernetics Society (Hong Kong Chapter). He is the recipient of The Hong Kong Polytechnic University President's Award for Excellent Achievement.

Chun-Hung Chiu is a Research Associate in the Department of Management Sciences at the City University of Hong Kong. He received his Ph.D. degree in Systems Engineering and Engineering Management from the Chinese University of Hong Kong, Hong Kong, in 2004. His research interests include supply chain management, operations management, and mathematical finance. Recently, his research focuses on the supply chain coordination with price/sales effort-dependent demand, operations management in socioeconomic problems, and effects of dynamic trading on portfolio management. He has published in journals such as *IEEE Transactions*, *Automatica*, *Quantitative Finance*, and *Production and Operations Management*.

Chapter 1
Mean-Risk Analysis: An Introduction

1.1 Background of Supply Chain Risk Analysis

Risk is a term commonly used in business. In a supply chain, risk is associated with the decision making problems whenever there exist uncertain outcomes and some of the outcomes are unfavorable (Lowrance 1976; Haimes 2004; Tuncel and Alpan 2010). There are different perspectives of supply chain risk analysis and management (Tang 2006; Sodhi et al. 2012[1]). One of these perspectives focuses on examining the *supply chain disruption risk* which is defined as the supply chain risk associated with anthropogenic or natural disruptions such as earthquakes, hurricanes, terrorist attacks, big storms, economic crises, diseases, etc. Another perspective links risk in the supply chain with some "normal uncertainties" such as demand volatility, supply uncertainty, and cost–revenue variation; this kind of risk is termed as the *supply chain operational risk*[2], and it is the major area of supply chain risk that we examine in this book.

Traditionally, in many supply chain management problems such as inventory control, risk associated with inventory decisions (derived from the probable mismatch between demand and supply inventory) is quantified by the "expected-measure-of-risk" such as expected overstocking and understocking costs. Despite being computationally tractable and easy to estimate, the use of expected measure alone obviously lacks precision. It is easy to find an example in inventory problems in which there exist two scenarios with the "same level of risk as measured by the expected overstocking and understocking inventory costs." However, the probability

[1] It is known that there is very diverse perception towards the field of "supply chain risk management." See Sodhi et al. (2012) for some consolidated recent views and findings from researchers in this emerging area.

[2] The interested reader is referred to Tang (2006) and Zsidisin and Ritchie (2008) for more comprehensive views on different perspectives on supply chain risks. For the review on operational hedging, refer to Van Mieghem (2003), Boyabath and Toktay (2004), Gaur and Seshadri (2005), and Van Mieghem (2007).

T.-M. Choi and C.-H. Chiu, *Risk Analysis in Stochastic Supply Chains: A Mean-Risk Approach*, International Series in Operations Research & Management Science 178, DOI 10.1007/978-1-4614-3869-4_1, © Springer Science+Business Media New York 2012

of yielding that "level of risk (under expected measure)" can be very different between the two scenarios because of the corresponding dispersion and variation of the random inventory overstocking and understocking costs. As such, additional measures and considerations should be incorporated into the analytical model for a more precise risk analysis. In addition, it is known that decision makers have different preferences or attitudes towards risk. In particular, risk aversion is a generally assumed risk attitude for decision makers in business because being conservative towards risk, and hence risk-averse, is crucial for the survival of many companies in the business world. As a result, how the degree of risk aversion affects the optimal decisions in supply chain operations is an important research issue which cannot be addressed using solely the "expected-measure-of-risk."

In light of the shortcomings and insufficiency of the "expected-measure" for risk analysis in supply chain management, alternative optimization objective measures and risk analysis models are proposed in the recent supply chain management literature. We review some of them as follows.[3]

von Neumann–Morgenstern Utility Functions. It is well known that in financial decision, one can model the respective optimization objective of a risk-averse decision maker by a von Neumann–Morgenstern utility function (of wealth in dollars, such as profit). To be specific, *a decision maker is risk-averse if and only if its von Neumann-Morgenstern utility function of wealth is strictly concave* (see Theorem 4, Ingersoll 1987, p. 37). Popular examples for risk-averse von Neumann–Morgenstern utility functions include the following: quadratic utility function, hyperbolic absolute risk aversion (HARA) utility function, and the linear risk tolerance (LRT) utility function family which consists of the (negative) exponential utility function, the isoelastic utility function, and the logarithmic utility function. In supply chain inventory management, a lot of studies have employed the von Neumann–Morgenstern utility function approach. For example, Atkinson (1979) studies the incentive issue in the newsvendor problem via the von Neumann–Morgenstern utility function approach. He finds that a risk-averse manager will order a smaller quantity compared to a risk-neutral manager. He also proposes a delegation scheme which can improve the situation. In his work that introduces various optimization objectives for the newsvendor problem, Lau (1980) discusses maximization of the expected utility and finds that the problem is complicated and the optimal solution can only be determined via a numerical method. Bouakiz and Sobel (1992) study the multiperiod inventory replenishment policy with the goal of minimizing the expected utility of the net present value of cost. They examine both the finite and infinite horizons cases. They find that a base-stock policy is optimal when the ordering cost is linear. Later on, Eeckhoudt et al.

[3] Notice that the following review only covers a few key related areas and is not meant to be an exhaustive one. For example, the coherent measures of risk and the loss aversion issues are not reviewed here. Interested reader can refer to Artzner et al. (1999), Choi and Ruszczynski (2008) and Choi (2012) for more details on the coherent measures of risk, and Wang and Webster (2009) and Wang (2010) for more discussions on the inventory problems related to loss aversion.

(1995) model risk-averse ordering decision in the newsvendor problem by using the exponential utility function. Risk is measured by the expected utility. They study and present the comparative-static effects of changes with respect to various parameters of the problem. Keren and Pliskin (2006) derive the optimality conditions for the risk-averse newsvendor problem under expected utility maximization framework. They employ these conditions to solve the case when the newsvendor's utility function is any increasing differentiable function and the demand follows a uniform distribution. They discuss the properties of the optimal solution and propose how it may be used for assessing the utility functions. Chen et al. (2007) propose a framework for incorporating risk aversion into a multiperiod inventory model. They show that the structure of the optimal dynamic programming policy for a decision maker with exponential utility function is very similar to the structure of the optimal policies under the risk-neutral case. They also extend the case to consider the scenario when both price and ordering quantity are decision variables of the optimal policy and get similar conclusion. Tapiero and Kogan (2009) examine the situation in which a retailer has to place an order before the product price is certain. They employ the utility function approach to capture the retailer's risk-averse preference. They show that the retailer's degree of risk aversion would affect its subjective assessment of future prices, which in turn also induces biases in its ordering decisions. Wang et al. (2009) investigate an interesting problem on whether a risk-averse newsvendor would order less at a higher selling price. They model the problem by the expected utility approach. They analytically show that under some fairly general conditions, the optimal order quantity of a risk-averse newsvendor decreases as the selling price increases. Choi and Ruszczynski (2011) study a multiitem newsvendor problem with an exponential utility function. They derive the monotonicity of the risk aversion's impact on the optimal decision. They prove that a closed-form approximated optimal solution exists when the ratio of the level of risk aversion to the number of items is sufficiently small and the product demands are all independent. They further examine the case when the product demands are correlated. By numerical examples, they show that increased risk aversion does not necessarily lead to lower optimal ordering quantities when product demands are sufficiently negatively correlated.

In the context of multiechelon supply chain system, Xie et al. (2011) study the optimal quality investment and pricing decision of a make-to-order supply chain. They model the risk-averse objective functions for supply chain agents as well as the integrated supply chain by the respective expected utility functions. They explore three different supply chain scenarios, namely the vertically integrated, the supplier-led, and the manufacturer-led scenarios. They find that compared with the risk-neutral case, the risk-averse supply chain possesses the same, a lower, and a higher product quality under the manufacturer-led, the vertically integrated, and the supplier-led scenarios, respectively. Giri (2011) studies a single period supply chain with a risk-averse retailer and two suppliers. He considers the case when the first supplier (called primary supplier) is cheaper but less reliable in terms of supply yield, whereas the second supplier (secondary supplier) is more expensive, perfectly reliable but its capacity is limited. He employs the exponential utility function to model the risk aversion of the retailer. He numerically demonstrates how the optimal inventory decisions under the risk-averse case are different from the risk-neutral case. He also conducts sensitivity analysis to reveal how the optimal decisions are affected by different cost–revenue parameters.

Despite being precise and theoretically sound, the von Neumann–Morgenstern utility function approach has limited real-world applications because of the huge difficulty in assessing a closed-form expression of the required utility function for individual decision maker in reality. As such, other approaches that carry good physical meanings and are easily implementable, such as many of those applied in finance, are adopted in practice.

Profit Target Probability Measures. The use of probability related measures is popular in supply chain inventory management. For example, Lau (1980) examines the single-period inventory optimization problem with the goal of maximizing the probability to achieve a certain predetermined fixed profit level. In particular, he considers two scenarios, one with the loss of goodwill kind of opportunity stockout cost and one without. For the scenario without this loss of goodwill stockout cost, he develops a closed-form expression for the optimal solution. For the scenario with this loss of goodwill stockout cost, he invents new numerical searching heuristics to solve the problem. Later on, Sankarasubramanian and Kumaraswamy (1983) study a similar problem and they focus on deriving insights based on the cases with exponential and uniform demand distributions. Instead of having a fixed profit target, Thakkar et al. (1983) extend the above problems by considering maximization of the probability of achieving a target return on investment. They derive the necessary optimality conditions, and prove the existence and the uniqueness of the optimal solution to the problem. Parlar and Weng (2003) consider an extended version of the original probability maximization problem introduced by Lau (1980) with a change of the objective function. To be specific, Parlar and Weng study the problem with the objective of maximizing the probability of exceeding the expected profit. Since the expected profit is a function of stocking quantity and is not a constant, it becomes a moving target. With this revised objective, Parlar and Weng propose a new method which combines the conventional expected profit maximization objective with the probability maximization objective to solve the problem. Shi et al. (2011) investigate the inventory competition problem in the newsvendor setting. They explore the problem with the objective of profit satisficing which is defined as the objective of maximizing the probability of achieving a profit target. They separate the analysis into two cases. In the first case, they consider the situation when each newsvendor's demand does not depend on its own stocking level and only depends on the stocking levels of all other competing newsvendors. In the second case, each newsvendor's demand depends on its own stocking level as well as the stocking levels of all other competing newsvendors. They discuss and prove the existence of the Nash equilibrium for each case. Most recently, Shi and Guo (2012) analytically study the case when a company owning multiple divisions has set a profit target for every division. They first derive the optimal profit target in closed form for every division when the company's optimization objective is to maximize its own expected profit. Afterwards, they examine the case when the company has an alternative optimization objective of maximizing the profit probability under a profit target setting. They find that for the case with "fair target setting," the optimal assigned profit target for the division which has a relatively high production cost is decreasing in its price elasticity. For the case if the headquarters of the company is in full control of the two homogeneous divisions, they find that each division's optimal profit target will be 50 % of the company's profit target when the price elasticity is two or larger than two. In the scope of supply chain coordination, Shi and Chen (2007) investigate a single-manufacturer single-retailer two-echelon supply chain with a newsvendor-type of product. They consider the case when both supply chain agents take the objective function of maximizing their own profit probabilities. They find an interesting result that a properly designed pure wholesale pricing contract can achieve supply chain coordination. Chen and Yano (2010) consider the use of rebate contract in a two-echelon supply chain with a seasonal product. The product's demand is weather sensitive. They study how the manufacturer can offer a weather-linked rebate contract to the retailer in a decentralized setting. They demonstrate that the weather-linked rebate contract that allows the manufacturer to achieve Pareto improvement can take different forms. They quantify risk for individual supply chain agents by employing the measure on "probability of having a lower profit in the presence of rebate." They further reveal that the manufacturer can fully hedge it risk of offering the weather-linked rebate contract by paying a risk premium.

Notice that the above profit probability related measures are easy to understand and widely implemented in practice. However, the degree of risk aversion of the decision maker is not clearly shown in these models and the probability measure alone does not tell the full picture associated with risk (e.g., the existence of a "low chance but huge loss" outcome).

Value-at-Risk (VaR) and Conditional Value-at-Risk (CVaR). The VaR analysis is popular in finance and has been applied in inventory management in recent years. For example, the VaR approach has been applied in several inventory control models (e.g., see the pioneering works Luciano et al. 2003 and Tapiero 2005) and it has also been considered in the multiproduct newsvendor (newsstand) problem. In particular, Özler et al. (2009) study a single-period multiitem stochastic inventory control problem with a VaR constraint. They adopt a mathematical programming approach to determine the optimal solution. They find that the proposed algorithm is robust enough to help solve the problems with correlated demands. Chiu and Choi (2010) examine the price-dependent newsvendor problem with a VaR objective. They study both the linear and multiplicative price-dependent demand distributions cases and analytically derive the optimal solutions for the problem under a VaR objective. Most recently, Chiu et al. (2011c) extend Chiu and Choi (2010) for the application of VaR newsvendor model in fashion retailing. They illustrate the model by both analytical and numerical approaches. Some counter-intuitive findings in supply chain coordination, such as the pure wholesale pricing contract outperforms the buyback contract, are reported. Despite being a popular measure, VaR suffers some theoretical flaws and it is not a coherent risk measure for nonelliptic distributions (Artzner et al. 1999). As a result, a better measure which is a coherent risk measure known as CVaR arises. In Wu et al. (2006), the concept of CVaR is employed to study an option-based "pay-to-delay" supply contract. They construct the dynamic optimization problem for the manufacturer and derive the optimal policy. They reveal how the degree of risk aversion affects the manufacturer's optimal strategy numerically. Gotoh and Takano (2007) investigate the newsvendor problem with CVaR objective. They demonstrate that the newsvendor problem with CVaR objective is tractable and the problem exhibits nice structural properties such as being convex. They develop linear programming formulation for the problem and reveal the efficiency of the linear programming solution method via numerical examples. Cheng et al. (2009) explore the bilevel newsvendor model in two-echelon supply chain systems. They model the retailer's objective as a CVaR objective and derive an analytical solution when the product demand follows a uniform distribution. Compared to the CVaR objective, they find that the retailer's CVaR is smaller when its objective is on maximizing expected profit. They also conduct sensitivity analysis to study the impacts brought by different model parameters on the optimal wholesale price and order quantity. Chen et al. (2009) further examine the CVaR objective as the decision criterion in the newsvendor problem with price-dependent demands. They examine the optimal pricing and stocking decisions and derive sufficient conditions for the existence of unique solution. They further reveal the neat monotonicity properties associated with the optimal pricing and ordering decisions. Chahar and Taaffe (2009) study the demand selection problem with all-or-nothing orders. They formulate the decision making problem by the CVaR approach and examine the optimal procurement policy. They compare both the risk-averse and risk-neutral cases and discuss how the all-or-nothing demands affect the expected profit and the frequency of procurement. Hsieh and Lu (2010) explore the return policy in a two-echelon single-period supply chain with one manufacturer and two risk-averse retailers with price-sensitive random demand. They characterize each retailer's risk-averse objective function by the CVaR approach. They develop manufacturer-led Stackelberg games in two scenarios where scenario one considers horizontal price competition between the retailers and scenario two does not. The effects brought by the degrees of risk aversion on the manufacturer's optimal return policy are revealed. Wu et al. (2010) examine the impacts of risk aversion on the optimal decisions in supply contract. They construct analytical optimization model for the manufacturer under the CVaR approach and study how the degree of risk aversion influences the manufacturer's optimal decisions. They reveal via numerical analysis the relationship between the manufacturer's degree of risk aversion and its optimal decisions. They further show the dependence of the decision variables on the price and cost parameters. Zhang et al. (2009) study the single and multiperiod inventory control models with risk-averse constraints. They employ both the VaR and the CVaR as risk measures. They propose a sample average approximation method to solve the stochastic optimization problem. They further present some numerical examples to demonstrate the applicability of the algorithm. Ma et al. (2010) analyze a single-period two-echelon supply chain with one manufacturer and one retailer. They consider the case when the retailer is risk-averse and the manufacturer is risk-neutral. They model the degree of risk aversion by the CVaR approach and construct a Nash-bargaining model in which the manufacturer and the retailer negotiate about the wholesale price and order quantity. They analytically show that there exists a unique Nash-bargaining equilibrium for both the endogenous and exogenous price cases. They reveal one interesting finding that even for the case when the retailer and the manufacturer have equal bargaining power, the retailer's bargaining power for the supply chain profit will increase if it is more risk-averse. Borgonovo and Peccati (2011) study the sensitivity analysis of inventory models. They employ the CVaR objective function for a case study and generate important insights by three different sensitivity analysis settings. Caliskan-Demirag and Chen (2011) study the rebates contracts (both consumer rebates and channel rebates) in the supply chain with a risk-averse retailer. They model the retailer's risk aversion by adopting the CVaR objective. Under a stochastic price dependent demand framework, they analyze the manufacturer's equilibrium contracting decisions and the retailer's equilibrium joint inventory and pricing decisions in a Stackelberg game framework. They show the monotonicity of the retailer's equilibrium decisions in its degree of risk aversion. They analytically explore the existence of the equilibrium for both formats of rebates. For a recent review and discussion on the use of VaR and CVaR objectives and the mean-CVaR rule for the newsvendor problem, please refer to Jammernegg and Kischka (2012).[4] Notice that the CVaR approach is a coherent risk measure which is consistent with, e.g., the rules of stochastic dominance. However, both VaR and CVaR are relatively advanced in the sense that the mathematics involved are not easily understood by industrialists and managers who are "laymen." As such, there is a need for another performance measure which is not only as useful as VaR and CVaR objectives, but also intuitive and easy to apply. This partially leads to the proposal of employing the mean-risk model which we will explore in Sect. 1.2.

1.2 Mean-Risk Analysis

Pioneered by Nobel laureate Professor Harry Markowitz in the 1950s, the mean-risk formulation (which includes both the classical mean-variance and mean-downside-risk approaches) is a fundamental and influential theory for risk management in portfolio investment (Markowitz 1959). In recent years, it has been applied for the analysis of different supply chain management problems. We review some representative works as follows.

[4] Interested reader is also referred to Borgonovo and Peccati (2009) for a comparison among various risk averse inventory models.

In the single-echelon setting, the seminal research which employs the use of a mean-risk objective in modeling the single period single echelon supply chain inventory problem appears in a section of Lau (1980) in which the newsvendor problem with a mean-standard-deviation (expected profit and standard-deviation of profit) objective is studied. After that, Choi et al. (1999) explore via the mean-variance approach the newsvendor problem with the safety-first objective. They consider the case when there exists the loss of goodwill opportunity cost for stockout. They find that the resulting optimization problem has a complicated structure and analytical closed-form solution does not exist. They thus demonstrate how a numerical searching method can be employed to solve the problem numerically. Choi et al. (2001) extend Choi et al. (1999) to consider the risk-averse newsvendor problem under a general mean-variance framework for both the cases with and without the loss of goodwill stockout cost. They show that the optimal ordering quantity can exceed the expected profit maximizing quantity when the loss of goodwill stockout cost is big. At the same period of time, Chen and Federgruen (2000) conduct a mean-variance analysis of various basic inventory models. They model a quadratic utility function for the inventory manager and then construct an efficient frontier for the noninferior solution points. They investigate and compare the "profit" and "cost" models. They also analyze the base stock policy with Poisson customer arrival and (R, nQ) model for the periodic-review inventory problem. Choi et al. (2008b) extend Lau (1980) and Choi et al. (2001) and consider the scenarios with risk-averse, risk-neutral, and risk-seeking decision makers for the newsvendor problem under different mean-variance frameworks. They analytically explore the optimal solution and derive the efficient frontier for each case. They also compare the cases with and without the loss of goodwill stockout opportunity cost. In the presence of the loss of goodwill stockout opportunity cost, Wu et al. (2009) analytically prove that when demand follows a power distribution, the newsvendor problem's optimal stocking quantity under a mean-variance framework can exceed the one under the risk-neutral case. Vaagen and Wallace (2008) conduct mean-semivariance analysis on the assortment planning in fashion supply chains. They develop the optimization model and derive the optimal portfolio based on the concept of hedging with the consideration of demand correlations. They show that building hedging portfolios, among the multiple competing items, is necessary for achieving optimality. They further reveal how mis-specifying the distributions can lead to improper hedging and finally yield very negative outcomes. In Choi et al. (2011), a multiperiod mean-variance model for the periodic review inventory policy is examined. Owing to the nonseparability of the variance of profit in the sense of dynamic programming in the original problem, a separable auxiliary problem is constructed. They analytically find the sufficient conditions under which the solutions for the auxiliary problem and the original problem converge. As a result, when the sufficient conditions are satisfied, the challenging multiperiod mean-variance inventory control problem can be solved by indirectly solving the auxiliary problem. Liu et al. (2011) employ the mean-variance objective to investigate the mass customization problem in which the risk-averse manufacturer makes three joint optimal decisions on retail pricing, consumer returns, and product modularity level. By comparing with the case when the manufacturer is risk-neutral, it is found that not only the optimal decisions are changed, the impacts brought by different given parameters are also significantly different. Liu and Nagurney (2011) utilize the mean-variance framework to analyze the supply chain offshore outsourcing problem under exchange rate risk and competition. They consider the presence of multiple companies that sell partially substitutable products in the market and have the choices of going outsourcing for production or manufacturing the goods in-house. They develop a variational inequality model which considers the pricing, transportation, outsourcing and in-house manufacturing decisions of the companies under exchange rate uncertainty. By employing extensive simulation analysis, they examine the level of profit and risk of the companies and reveal how the companies' degrees of risk aversion affect the optimal outsourcing decisions. Choi and Chiu (2012) study the newsvendor problem under the mean-variance and the mean-downside-risk frameworks. They relate the problems with different sustainability measures for the inventory problem. Most recently, Choi et al. (2012) extend Liu et al. (2011) to study the situation when demand and returns are correlated. By fitting into the industrial practice of fashion mass customization programs, they examine the conditions under which the strategy of adopting returns with full refund is a better strategy than no returns.

In addition to the single echelon mean-risk problems reviewed above, the multiechelon supply chain coordination problems are also well-studied under the mean-risk framework. The first piece of work which analyzes the supply chain in a multiechelon setting is Lau and Lau (1999). To be specific, Lau and Lau examine a two-echelon supply chain with a single manufacturer and single retailer with a returns policy. They consider the situation when both supply chain agents are risk-averse and possess a linear objective function of mean and variance. By focusing on the normally distributed demand, they generate many insights related to the impacts brought by the degrees of risk aversion on supply chain coordination via extensive numerical analysis.

Agrawal and Seshadri (2000) conduct the mean-variance analysis to a two-echelon supply chain with multiple risk-averse retailers. They model the mean-variance objective for each risk-averse retailer and propose the use of a menu of supply contracts to coordinate the respective supply chain. Considering a similar problem as Agrawal and Seshadri (2000)'s but in the continuous domain (with infinite number of retailers and their coefficients of risk aversion follow a continuous distribution), Chen and Seshadri (2005) employ the optimal control theory to solve the supply chain coordination problem. They analytically demonstrate that the distribution of the coefficients of risk aversion is critical and affects the supply chain structure significantly. They prove that the optimal menu of supply contracts is similar in structure to the one developed by Agrawal and Seshadri (2000). Similar to Lau and Lau (1999), Tsay (2002) also applies the mean-variance objective in analyzing a two-echelon supply chain. He analytically studies how risk sensitivity affects both sides of the supplier–retailer relationship under different scenarios of strategic power in the presence of the returns policy. He concludes by commenting on the importance of incorporating risk sensitivity into the planning of the optimal supply chain. After that, Gan et al. (2004, 2005) investigate the supply chain coordination challenge in a two-echelon supply chain with risk-averse agents. They employ the mean-risk approach and generate many important analytical results. Martinez-de-Albeniz and Simchi-Levi (2006) employ the mean-variance approach to study the trade-offs faced by a manufacturer which signs a portfolio of long-term contracts with its supplier. They consider the case when there exists a spot market and hence the manufacturer faces inventory risk if it purchases too many contracts and the spot price risk if it purchases too little. They further derive the set of mean-variance efficient portfolios that the manufacturer can hold to achieve the mean-variance dominating pairs. Choi et al. (2008a) extend Lau and Lau (1999) and analytically study the returns policy for supply chain coordination under a mean-variance framework. They examine various scenarios which include the cases with a centralized supply chain and a decentralized supply chain. They also study the situation when there is information asymmetry on the degree of risk aversion in the supply chain. Choi et al. (2008c) study the two-echelon supply chain coordination problem when the agents take different/same risk attitudes (which can be risk-seeking, risk-averse, or risk-neutral) under a mean-variance framework. They find that the achievability of supply chain coordination depends on how different the risk related thresholds between the two supply chain agents are. In Wei and Choi (2010), inspired by the observed industrial practice on wholesale pricing and profit sharing policy, they conduct a mean-variance analysis for this contract on a two-echelon supply chain. They establish the analytical closed form necessary and sufficient conditions for coordinating the supply chain by a wholesale pricing and profit sharing scheme under an information symmetric case. They then study the case with information asymmetry and show that the retailer can be beneficial by pretending to be more risk-averse. They propose an innovative measure, with the use of minimum quantity commitment, for the manufacturer to impose on the supply contract so as to avoid the retailer's cheating case from happening. Choi (2011) analytically studies the use of RFID under vendor-managed inventory (VMI) policy in a two-echelon single-manufacturer single-retailer supply chain. He constructs the supply chain models for a multiperiod retail replenishment problem with and without RFID under the mean-variance framework. He then proposes measures to achieve supply chain coordination with the use of RFID. He analytically illustrates that if the RFID tag cost is sufficiently small, employing RFID technology will lead to an improved supply chain with both a larger expected profit and a smaller level of risk (as quantified by the variance of profit). He also reveals that it is beneficial for the manufacturer to take the initiative to share the retailer's RFID implementation expense because it will help coordinate the supply chain and also reduce the manufacturer's level of risk. Hung et al. (2011) study a two-stage risk-averse newsvendor model in which upstream suppliers have short lead-time capacity to produce products for the retailers. They employ the mean-variance model to study how the inventory and supply risks of the retailers can be pooled and shared among different supply chains by treating capacity as commodities and trading them to hedge risks. They show that their proposed new hedging mechanism can efficiently manage inventory. Chiu et al. (2011a) conduct a mean-variance analysis of a supply chain under target sales rebate contract. They illustrate how a target sales rebate contract can coordinate the supply chain in different scenarios. Shen et al. (2012) investigate the markdown policy in a supply chain with a fashion product. They focus on the case when the upstream manufacturer is risk-averse and possesses a mean-variance objective function. They analytically establish the conditions for achieving supply chain coordination. They employ real data collected from the industry to examine how the supply chain and its agents perform with respect to the contract and market parameters under the existing industrial practice. They develop interesting insights related to the profit's coefficient of variation in the supply chain. Most recently, by quantifying risk by the variance of profit, Jornsten et al. (2012) study how real options can be used to transfer risk between the manufacturer and the retailer in the supply chain when demand follows a discrete distribution. They formulate the contracting model in the Pareto-optimal setting. They consider that the manufacturer's objective is to design "feasible" real option contracts with which both the retailer and the manufacturer would enjoy at least as much expected profit as in the original contract. They find that the set of feasible contracts is a complicated nonconvex set.

As a remark, the mean-risk analysis is also applied to explore supply chain management problems with information updating. The first piece of work which employs mean-risk analysis for studying a supply chain with information updating is Choi et al. (2003). To be specific, Choi et al. (2003) construct a two-stage two-ordering dynamic optimization model with Bayesian information updating. They utilize the variance of profit for studying the level of profit uncertainty (and hence risk) associated with each policy under study. Afterwards, Choi et al. (2004) extend the information updating model in Choi et al. (2003) from two stages to N stages (where $N > 2$). They consider the situation when there is only one ordering opportunity. By formulating the problem as an optimal-stopping time model, an optimal stocking policy is developed. In addition, by using the mean-variance analysis, they find that the level of risk associated with the ordering decision is decreasing with the ordering time point. Choi and Chow (2008) conduct an extensive mean-variance analysis for a two-echelon supply chain under quick response program (with information updating). They illustrate how the commonly adopted measures such as price commitment policy, service commitment policy, and returns policy can be properly set in order to achieve the win–win situation under which the supply chain agents will all be better off in both expected profit and risk. They conduct extensive numerical analysis and derive analytical conditions for achieving supply chain coordination. Most recently, Buzacott et al. (2011) apply the mean-variance approach to look into a class of commitment–option supply contracts. They explore the problem's structural properties and demonstrate how a mean-variance trade-off with information update can be carried out. They further illustrate how the supply contract setting with risk consideration would differ from the risk-neutral case.

Table 1.1 presents a summary of the related literature on supply chain risk analysis.

Table 1.1 The related literature on supply chain risk analysis

	Single echelon		
	Single period	Multiple periods	Multi echelons
von Neumann–Morgenstern utility functions	Atkinson (1979), Lau (1980), Eeckhoudt et al. (1995), Keren and Pliskin (2006), Tapiero and Kogan (2009), Wang et al. (2009), Choi and Ruszczynski (2011)	Bouakiz and Sobel (1992), Chen et al. (2007)	Giri (2011), Xie et al. (2011)
Profit target probability measures	Lau (1980), Sankarasubramanian and Kumaraswamy (1983), Parlar and Weng (2003), Shi and Guo (2012)		Shi and Chen (2007), Chen and Yano (2010), Shi et al. (2011)
VaR, CVaR	Tapiero (2005), Gotoh and Takano (2007), Chen et al. (2009), Özler et al. (2009), Chahar and Taaffe (2009), Chiu and Choi (2010), Borgonovo and Peccati (2011), Jammernegg and Kischka (2012)	Luciano et al. (2003), Tapiero (2005), Zhang et al. (2009)	Wu et al. (2006), Cheng et al. (2009), Hsieh and Lu (2010), Ma et al. (2010), Wu et al. (2010), Caliskan-Demirag and Chen (2011), Chiu et al. (2011c)
Mean-risk (mean-variance, mean-semivariance)	Lau (1980), Choi et al. (1999, 2001), Chen and Federgruen (2000), Vaagen and Wallace (2008), Choi et al. (2008b), Wu et al. (2009), Liu et al. (2011), Liu and Nagurney (2011), Choi and Chiu (2012)	Choi et al. (2011)	Lau and Lau (1999), Agrawal and Seshadri (2000), Tsay (2002), Gan et al. (2004, 2005), Chen and Seshadri (2005), Martinez-de-Albeniz and Simchi-Levi (2006), Choi et al. (2008a, c), Wei and Choi (2010), Buzacott et al. (2011), Choi (2011), Chiu, Choi and Li (2011), Hung et al. (2011), Jornsten et al. (2012), Shen et al. (2012)

Undoubtedly, the mean-risk approach has already been shown to be useful in conducting risk analysis in stochastic supply chain models. In the next section, we discuss the pros and cons of employing the mean-risk approach for conducting supply chain risk analysis.

1.3 Why Mean-Risk?

As we discussed in the earlier sections, the mean-risk formulation is a crucial theory for risk management in finance (Markowitz 1959). In fact, in the literature, the mean-risk approach and the von Neumann–Morgenstern utility function approach are two most well developed methods for studying decision making problems under risk. When we compare between them, we understand that the von Neumann–Morgenstern utility function approach is indeed more precise in general, but its real-world application is limited owing to the difficulty in getting a closed-form expression of the utility function for individual decision maker. It is also not necessarily intuitive to decision makers such as managers in the industry. On the contrary, the mean-risk approach aims at providing a useful, implementable (because only two statistics: the mean and the risk measure such as the "variance" or the "downside-risk-measure" are needed), intuitive (easy to understand), and approximate solution (see Van Mieghem 2003; Choi et al. 2008a). As a remark, it is well known that the mean-variance objective is equivalent to a quadratic von Neumann–Morgenstern utility function. Moreover, even though a general form of von Neumann–Morgenstern utility function cannot be expressed fully in terms of mean and another simple risk measure such as variance only, as derived by in the literature (see Van Mieghem 2003), optimizing a von Neumann–Morgenstern utility function with a constant coefficient of risk aversion is equivalent to maximizing a mean-variance performance measure. There are also other findings reported in the literature which strongly demonstrate that the mean-risk approach can give a solution close to the global optimum under the von Neumann–Morgenstern utility function approach (see, e.g., Levy and Markowitz 1979).

As a remark, as the original version of the mean-risk model, the Markowitz' mean-variance approach has a major theoretical flaw because both the upside and downside deviations from the mean (e.g., the expected profit) are included in the calculation of variance and hence risk. Thus, even the deviation on "upside" (i.e., bigger than the "mean," e.g., having better profit than the expected profit) is treated as something bad and will be classified as a part of risk. Although this problem vanishes naturally in Markowitz's original portfolio management model when the underlying distribution of returns is normal (symmetric), it is not trivial in most supply chain management problems such as inventory control. As a consequence, an alternative risk measure known as downside-risk measure (such as semideviation (Ogryczak and Ruszczynski 2001)) hence arises. The downside risk measure is actually similar to the variance measure, but we discard the upside deviation when we calculate the risk measure. Here, the upside deviation is usually defined with respect to either a constant threshold (e.g., a profit target) or the mean (e.g., the expected profit). Notice that there are some studies devoted to revealing the differences and similarities between the mean-variance and mean-downside-risk approaches, e.g., Grootveld and Hallerbach (1999) and Choi and Chiu (2012). A rather common finding is this: In most cases, the mean-variance approach would yield similar results as the mean-downside risk approach, and no huge difference exists in most practical cases. As such, we argue that for many applications, one can apply either model in supply chain analysis in practice.

In addition, according to Haimes (2004), the criteria for good risk analysis measure include the following:

1. Comprehensive
2. Adherent to evidence
3. Logically sound
4. Practical
5. Open to evaluation
6. Based on explicit assumptions and premises
7. Compatible with institutions
8. Conducive to learning
9. Attuned to risk communication
10. Innovative

The mean-risk approach possesses nice features which basically satisfy these criteria. As such, it is a good approach for conducting risk analysis in a supply chain.

1.4 Organization

The rest of this book is organized as follows.

Chapter 2 is devoted to the mean-risk analysis for the single period inventory problem. We construct the basic single period inventory control models under the mean-variance and mean-semideviation frameworks. We review and explore the expected profit, the variance of profit, and the semideviation of profit functions. We analytically show that the expected profit function is concave, and prove that both the variance of profit and the semideviation of profit functions are bounded above. We discuss the case with a normally distributed demand as an illustrative example. We then proceed to examine the general objective functions in the mean-variance and mean-semideviation domains and derive the efficient region. We investigate the efficient frontiers for both models. We further show that the mean-variance and the mean-semideviation models have the same solution when the two models are "fairly" compared and constructed. We then present numerical analyses and explore how the retail selling price and demand uncertainty affects the efficient frontier.

Chapter 3 conducts the mean-risk analysis for the multiperiod inventory problem. We choose the classical (*R*, *nQ*) multiperiod inventory replenishment model with an infinite horizon as an example to demonstrate how a mean-risk analysis can be carried out for a multiperiod inventory problem. We construct the analytical mean-risk model by taking the long run-average profit as the "mean," and propose the variance of on-hand inventory and the variance of one period profit as measures of "risk." We derive the closed-form analytical expressions of the long run-average profit, the variance of on-hand inventory and the variance of one period profit. We then discuss how to construct the efficient frontier in the mean-risk domain.

Chapter 4 examines supply chain coordination problem with mean-risk analysis. We consider a single-period two-echelon supply chain with a single risk-neutral manufacturer and a single risk-averse retailer. We model the risk aversion preference of the retailer by a mean-risk framework. We investigate how target sales rebate supply contract can help to coordinate the supply chain in the sense of maximizing the expected profit of the supply chain system. We further extend the mean-risk analysis to study the challenging coordination problem in which market demand is sales effort dependent.

Chapter 5 concludes this book with discussions on future research directions. We separate the discussions into four sections in which each section focuses on one important future research direction. To be specific, the first proposed future research direction in Chap. 5 is to expand the scope of analysis from expected profit/expected cost models to mean-risk models with two approaches. Approach one incorporates the mean-risk objective into the respective supply chain optimization problem directly. Approach two refers to a two-level analysis framework in which the first level still employs the expected measure as the optimization objective and the second level uses the risk measure to analyze the performance of the supply chain with respect to the first level's optimization decision. The second proposed future research direction in Chap. 5 is to study information asymmetric supply chains under the mean-risk domain. We argue that the mean-risk analysis is especially vital for the supply chain with information asymmetry because there are more sources of uncertainty. We discuss several important topics such as asymmetric information on demand distribution and cost–revenue parameters in the supply chain between supply chain agents, the unknown risk preference, and the moral hazard issue. The third proposed future research direction in Chap. 5 is related to conducting mean-risk analysis for more complex supply chain system. We specifically propose to study in longer supply chains the relationships between performance of channel leadership and risk aversion, in wider supply chains the coordination challenges by innovative menu of contracts and dynamic contracts, and in multiperiod supply chains the new optimization methods and coordination mechanism. The last proposed future research direction in Chap. 5 concerns conducting behavioral research. We propose to verify, refine, and extend the mean-risk analytical supply chain models and coordination problems by human-subject based behavioral experiments. We finally conclude by presenting a summary of these future research directions, together with the related references, in a table.

References

Agrawal, V., & Seshadri, S. (2000). Risk intermediation in supply chains. *IIE Transactions, 32*, 819–831.

Artzner, P., Delbaen, F., Eber, J., & Heath, D. (1999). Coherent measures of risk. *Mathematical Finance, 9*, 203–228.

Atkinson, A. A. (1979). Incentives, uncertainty, and risk in the newsboy problem. *Decision Science, 10*, 341–357.

Borgonovo, E., & Peccati, L. (2009). Financial management in inventory problems: Risk averse vs risk neutral policies. *International Journal of Production Economics, 118*, 233–242.

Borgonovo, E., & Peccati, L. (2011). Finite change comparative statics for risk-coherent inventories. *International Journal of Production Economics, 131*, 52–62.

Bouakiz, M., & Sobel, M. J. (1992). Inventory control with an exponential utility criterion. *Operations Research, 40*, 603–608.

Boyabath, O., & Toktay L. B. (2004). *Operational hedging: A review with extensions.* Working paper, INSEAD.

Buzacott, J., Yan, H., & Zhang, H. (2011). Risk analysis of commitment–option contracts with forecast updates. *IIE Transactions, 43*, 415–431.

Caliskan-Demirag, O., & Chen, Y. (2011). Customer and retailer rebates under risk aversion. *International Journal of Production Economics, 133*, 736–750.

Chahar, K., & Taaffe, K. (2009). Risk averse demand selection with all-or-nothing orders. *Omega, 37*, 996–1006.

Chen, F., & Federgruen, A. (2000). *Mean-variance analysis of basic inventory models.* Working paper, Columbia University.

Chen, Y. J., & Seshadri, S. (2005). Supply chain structure and demand risk. *Automatica, 42*, 1291–1299.

Chen, X., Sim, M., Simchi-Levi, S., & Sun, P. (2007). Risk aversion in inventory management. *Operations Research, 55*, 828–842.

Chen, Y. F., Xu, M., & Zhang, Z. (2009). A risk-averse newsvendor model with CVaR criterion. *Operations Research, 57*, 1040–1044.

Chen, Y. F., & Yano, C. A. (2010). Improving supply chain performance and managing risk under weather-related demand uncertainty. *Management Science, 56*, 1380–1397.

Cheng, L., Wan, Z., & Wang, G. (2009). Bilevel newsvendor models considering retailer with CVaR objective. *Computers and Industrial Engineering, 57*, 310–318.

Chiu, C. H., & Choi, T. M. (2010). Optimal pricing and stocking decisions for newsvendor problem with value-at-risk consideration. *IEEE Transactions on Systems, Man and Cybernetics, Part A, 40*, 1116–1119.

Chiu, C. H., Choi, T. M., Hao, G., & Li, X. (2012a). *Coordinating supply chains with a global-brand and multiple risk-averse retailers.* Working paper, The Hong Kong Polytechnic University.

Chiu, C. H., Choi, T. M., & Li, X. (2011°). Supply chain coordination with risk sensitive agent under target sales rebate contracts. *Automatica, 47*, 1617–1625.

Chiu, C. H., Choi, T. M., & Tang, C. S. (2011b). Price, rebate, and returns supply contracts for coordinating supply chains with price dependent demand. *Production and Operations Management, 20*, 81–91.

Chiu, C. H., Zheng, J., & Choi, T. M. (2011c). Optimal pricing and inventory decisions for fashion retailers under value-at-risk objective: Applications and review. In T.M. Choi (Ed.), *Fashion supply chain management: Industry and business analysis* (pp. 100–109). Hershey, PA: IGI Global.

Choi, T. M. (2011). Coordination and risk analysis of VMI supply chains with RFID technology. *IEEE Transactions on Industrial Informatics, 7*, 497–504.

Choi, S. (2012). A multi-product risk-averse newsvendor with law invariant coherent measures of risk. In T.M. Choi (Ed.), *Handbook on newsvendor problems: Models, extensions and applications*. New York: Springer.

Choi, T. M., & Chiu, C. H. (2012). Mean-downside-risk and mean-variance newsvendor models: Implications for sustainable fashion retailing. *International Journal of Production Economics, 135*, 552–560.

Choi, T. M., Chiu, C. H., & Fu, P. (2011). Periodic review multi-period inventory control under a mean-variance optimization objective. *IEEE Transactions on Systems, Man, and Cybernetics, Part A - Systems and Humans, 41*, 678–682.

Choi, T. M., & Chow, P. S. (2008). Mean-variance analysis of quick response program. *International Journal of Production Economics, 114*, 456–475.

Choi, T. M., Li, D., & Yan, H. (1999). *Newsvendor problem with the safety-first objective*. Working paper, The Chinese University of Hong Kong.

Choi, T. M., Li, D., & Yan, H. (2001). Newsvendor problem with mean-variance objectives. *Proceedings of the 5th International Conference on Optimization: Techniques and Applications* (Vol. 4, pp. 1860–1867).

Choi, T. M., Li, D., & Yan, H. (2003). Optimal two-stage ordering policy with Bayesian information updating. *Journal of the Operational Research Society, 54*, 846–859.

Choi, T. M., Li, D., & Yan, H. (2004). Optimal returns policy for supply chain with e-marketplace. *International Journal of Production Economics, 88*, 205–227.

Choi, T. M., Li, D., & Yan, H. (2008a). Mean-variance analysis of a single supplier and retailer supply chain under a returns policy. *European Journal of Operational Research, 184*, 356–376.

Choi, T. M., Li, D., & Yan, H. (2008b). Mean-variance analysis for the newsvendor problem. *IEEE Transactions on Systems, Man, and Cybernetics, Part A - Systems and Humans, 38*, 1169–1180.

Choi, T. M., Li, D., Yan, H., & Chiu, C. H. (2008c). Channel coordination in supply chains with members having mean-variance objectives. *Omega, 36*, 565–576.

Choi, T. M., Liu, N., Ng, F., & Yuen, M. (2012). *Mean-risk analysis of fashion mass customization programme with demand-return correlation*. Working paper, The Hong Kong Polytechnic University.

Choi, S., & Ruszczynski, A. (2008). A risk-averse newsvendor with law-invariant coherent measures of risk. *Operations Research Letters, 36*, 77–82.

Choi, S., & Ruszczynski, A. (2011). A multi-product risk-averse newsvendor with exponential utility function. *European Journal of Operational Research, 214*, 78–84.

Eeckhoudt, L., Gollier, C., & Schlesinger, H. (1995). The risk-averse (and prudent) newsboy. *Management Science, 41*, 786–794.

Gan, X., Sethi, S. P., & Yan, H. (2004). Coordination of supply chains with risk-averse agents. *Production and Operations Management, 13*, 135–147.

Gan, X., Sethi, S. P., & Yan, H. (2005). Channel coordination with a risk-neutral supplier and a downside-risk-averse retailer. *Production and Operations Management, 14*, 80–89.

Gaur, V., & Seshadri, S. (2005). Hedging inventory risk through market instruments. *Manufacturing and Service Operations Management, 7*, 103–120.

Giri, B. C. (2011). Managing inventory with two suppliers under yield uncertainty and risk aversion. *International Journal of Production Economics, 133*, 80–85.

Gotoh, J., & Takano, Y. (2007). Newsvendor solutions via conditional value-at-risk minimization. *European Journal of Operational Research, 179*, 80–96.

Grootveld, H., & Hallerbach, W. (1999). Variance vs. downside risk: Is there really that much difference? *European Journal of Operational Research, 114*, 304–319.

Haimes, Y. Y. (2004). *Risk modeling, assessment, and management* (2nd ed.). New York: Wiley-Interscience.

Hsieh, C. C., & Lu, Y. T. (2010). Manufacturer's return policy in a two-stage supply chain with two risk-averse retailers and random demand. *European Journal of Operational Research, 207*, 514–523.

Hung, Y. H., Li, L. Y. O., & Cheng, T. C. E. (2011). Transfer of newsvendor inventory and supply risks to sub-industry and the public by financial instruments. *International Journal of Production Economics*, in press.

Ingersoll, J. C. (1987). *Theory of financial decision making*. Lanham, MD: Rowman and Littefield Publishers.

Jammernegg, W., & Kischka, P. (2012). Newsvendor problems with VaR and CVaR consideration. In T.M. Choi (Ed.), *Handbook on newsvendor problems: Models, Extensions and Applications*. New York: Springer.

Jornsten, K., Nonas, S. L., Sandal, L., & Uboe, J. (2012). Transfer of risk in the newsvendor model with discrete demand. *Omega, 40*, 404–414.

Keren, B., & Pliskin, J. S. (2006). A benchmark solution for the risk-averse newsvendor problem. *European Journal of Operational Research, 174*, 1643–1650.

Lau, H. S. (1980). The newsboy problem under alternative optimization objectives. *Journal of the Operational Research Society, 31*, 525–535.

Lau, H. S., & Lau, A. H. L. (1999). Manufacturer's pricing strategy and returns policy for a single-period commodity. *European Journal of Operational Research, 116*, 291–304.

Levy, H., & Markowitz, H. M. (1979). Approximated expected utility by a function of mean and variance. *American Economics Review, 69*, 308–317.

Liu, N., Choi, T. M., Yuen, M. C. W., & Ng, F. (2011). Optimal pricing, modularity and return policy under mass customization. *IEEE Transactions on Systems, Man, and Cybernetics, Part A*. doi:10.1109/TSMCA.2011.2170063. published online.

Liu, Z., & Nagurney, A. (2011). Supply chain outsourcing under exchange rate risk and competition. *Omega, 39*, 539–549.

Lowrance, W. W. (1976). *Of acceptable risk*. Los Altos, CA: William Kaufmann.

Luciano, E., Peccati, L., & Cifarelli, D. M. (2003). VaR as a risk measure for multiperiod static inventory models. *International Journal of Production Economics, 81–82*, 375–384.

Ma, L., Liu, F., Li, S., & Yan, H. (2010). Channel bargaining with risk-averse retailer. *International Journal of Production Economics*, in press http://dx.doi.org/10.1016/j.ijpe.2010.08.016.

Markowitz, H. M. (1959). *Portfolio selection: Efficient diversification of investment*. New York: John Wiley & Sons.

Martinez-de-Albeniz, V., & Simchi-Levi, D. (2006). Mean-variance trade-offs in supply contracts. *Naval Research Logistics, 53*, 603–616.

Ogryczak, W., & Ruszczynski, A. (2001). On consistency of stochastic dominance and mean–semideviation models. *Mathematical Programming, Series B, 89*, 217–232.

Özler, A., Tan, B., & Karaesmen, F. (2009). Multi-product newsvendor problem with value-at-risk considerations. *International Journal of Production Economics, 117*, 244–255.

Parlar, M., & Weng, Z. K. (2003). Balancing desirable but conflicting objectives in the newsvendor problem. *IIE Transactions, 35*, 131–142.

Sankarasubramanian, E., & Kumaraswamy, S. (1983). Optimal ordering quantity for pre-determined Level of profit. *Management Science, 29*, 512–514.

Shen, B., Choi, T. M., Wang, Y., & Lo, C. K. Y. (2012). *Coordination of fashion supply chains with risk averse supplier by markdown money policy*. Working paper, The Hong Kong Polytechnic University.

Shi, C. V., & Chen, B. (2007). Pareto-optimal contracts for a supply chain with satisficing objectives. *Journal of the Operational Research Society, 58*, 751–760.

Shi, C. V., & Guo, L. (2012). Profit target setting for multiple divisions: A newsvendor perspective. In T.M. Choi (Ed.), *Handbook on newsvendor problems: Models, extensions and applications*. New York: Springer.

Shi, C. V., Yang, S., Xia, Y., & Zhao, X. (2011). Inventory competition for newsvendors under the objective of profit satisficing. *European Journal of Operational Research, 215*, 367–373.

Sodhi, M. S., Son, B. G., & Tang, C. S. (2012). Researchers' perspectives on supply chain risk management. *Production and Operations Management, 21*, 1–13.

Tang, C. S. (2006). Perspectives in supply chain risk management. *International Journal of Production Economics, 103*, 451–488.
Tapiero, C. S. (2005). Value at risk and inventory control. *European Journal of Operational Research, 163*, 769–775.
Tapiero, C. S., & Kogan, K. (2009). Risk-averse order policies with random prices in complete market and retailer's private information. *European Journal of Operational Research, 196*, 594–599.
Thakkar, R. B., Finley, D. R., & Liao, W. M. (1983). A stochastic demand CVP model with return on investment criterion. *Contemporary Accounting Research, 1*, 77–86.
Tsay, A. A. (2002). Risk sensitivity in distribution channel partnerships: Implications for manufacturer return policies. *Journal of Retailing, 78*, 147–160.
Tuncel, G., & Alpan, G. (2010). Risk assessment and management for supply chain networks: A case study. *Computers in Industry, 61*, 250–259.
Vaagen, H., & Wallace, S. W. (2008). Product variety arising from hedging in the fashion supply chains. *International Journal of Production Economics, 114*, 431–455.
Van Mieghem, J. A. (2003). Capacity management, investment, and hedging: Review and recent developments. *Manufacturing and Service Operations Management, 5*, 269–301.
Van Mieghem, J. A. (2007). Risk mitigation in newsvendor network: Resource diversification, flexibility, sharing, and hedging. *Management Science, 53*, 1269–1288.
Wang, C. X. (2010). The loss-averse newsvendor game. *International Journal of Production Economics, 124*, 448–452.
Wang, C. X., & Webster, S. (2009). The loss-averse newsvendor problem. *Omega, 37*, 93–105.
Wang, C. X., Webster, S., & Suresh, N. C. (2009). Would a risk-averse newsvendor order less at a higher selling price? *European Journal of Operational Research, 196*, 544–553.
Wei, Y., & Choi, T. M. (2010). Mean-variance analysis of supply chains under wholesale pricing and profit sharing scheme. *European Journal of Operational Research, 204*, 255–262.
Wu, J., Li, J., Wang, S., & Cheng, T. C. E. (2009). Mean–variance analysis of the newsvendor model with stockout cost. *Omega, 37*, 724–730.
Wu, J., Wang, S., Chao, X., Ng, C. T., & Cheng, T. C. E. (2010). Impact of risk aversion on optimal decisions in supply contracts. *International Journal of Production Economics, 128*, 569–576.
Wu, J., Yue, W., Yamamoto, Y., & Wang, S. (2006). Risk analysis of a pay to delay capacity reservation contract. *Optimization Methods and Software, 21*, 635–651.
Xie, G., Yue, W., Wang, S., & Lai, K. K. (2011). Quality investment and price decision in a risk-averse supply chain. *European Journal of Operational Research, 214*, 403–410.
Zhang, D., Xu, H., & Wu, Y. (2009). Single and multi-period optimal inventory control models with risk-averse constraints. *European Journal of Operational Research, 199*, 420–434.
Zsidisin, G. A., & Ritchie, B. (2008). *Supply chain risk*. New York: Springer.

Chapter 2
Mean-Risk Analysis of Single-Period Inventory Problems

In this chapter, we carry out mean-risk analysis for the single-period inventory problems. We first construct the basic inventory control model under the mean-risk framework. We then present two kinds of mean-risk objective functions and analytically prove the existence of an efficient region for either mean-risk model. After that, we explore the construction of the efficient frontier in the mean-risk domain. Before we conclude, numerical analysis is presented to illustrate the mean-risk trade-off in the single-period inventory decision making problems.

2.1 Basic Model Under Mean-Risk Framework

Among all the single-period inventory problems under demand uncertainty, the newsvendor problem is probably the most widely studied one. In the classical single-period single-item newsvendor problem, a retailer (known as newsvendor) orders a certain amount of perishable product (such as short-life fashionable items) from its supplier with a unit ordering cost c. The product is sold in the market with a unit retail selling price (revenue) p, where $p>c$ for a single selling season. In this chapter, the unsold product is assumed to have a zero salvage value (notice that the analysis will be similar and the insights are also similar when we consider a nonzero salvage value; we thus drop it in our analysis here for the sake of simplicity and neatness). The product's demand, x, is a continuous random variable which follows a certain stationary distribution with a probability density function $f(x)$ and cumulative distribution function $F(x)$. The mean of demand is represented by e and the variance is denoted by v. To avoid trivial and confusing cases, we consider in this chapter that the moments of x are finite and there is a one-to-one mapping between $F(x)$ and its argument. We further assume that the inverse function of $F(x)$ exists and denote it by $F^{-1}(x)$. Before the season starts, the retailer needs to

T.-M. Choi and C.-H. Chiu, *Risk Analysis in Stochastic Supply Chains: A Mean-Risk Approach*, International Series in Operations Research & Management Science 178,
DOI 10.1007/978-1-4614-3869-4_2, © Springer Science+Business Media New York 2012

determine the optimal order quantity for this fashionable product. Throughout this chapter, the order quantity is represented by q. Define:

$$\max(0, X) = (X)^+,$$
$$n(q) = \int_0^q F(x)\mathrm{d}x,$$

and

$$\lambda(q) = q - n(q).$$

Proposition 2.1 *(a) $n(q)$ is an increasing function of q.* (b) *$\lambda(q)$ is an increasing function of q.*

Proof of Proposition 2.1 (a) By differentiating $n(q)$ with respect to q, we have the following: $\mathrm{d}n(q)/\mathrm{d}q = F(q) \geq 0$. (b) Similarly, we have $\mathrm{d}\lambda(q)/\mathrm{d}q = 1 - F(q) \geq 0$.

Notice that the results of Proposition 2.1 are useful for us to derive important structural properties for various performance measures under mean-risk models.

With the above details, we can derive the profit function $P(q)$ as follows:

$$\begin{aligned} P(q) &= \min(x, q)p - cq \\ &= pq - p(q - x)^+ - cq \\ &= (p - c)q - p(q - x)^+ . \end{aligned} \tag{2.1}$$

With (2.1), by taking expectation, we get the expected profit $E[P(q)]$ below:

$$\begin{aligned} E[P(q)] &= (p - c)q - pE[(q - x)^+] \\ &= (p - c)q - p\int_0^q (q - x)\mathrm{d}F(x) . \\ &= (p - c)q - p\int_0^q F(x)\mathrm{d}x \end{aligned} \tag{2.2}$$

Similarly, we can derive the expressions for the variance of profit $V[P(q)]$ and the downside risk measure "semideviation of profit" $S_{\text{downside}}[P(q)]$ as follows:

$$\begin{aligned} V[P(q)] &= p^2 Var[(q - x)^+] \\ &= p^2\left(2qn(q) - 2\int_0^q xF(x)\mathrm{d}x - [n(q)]^2\right), \\ S_{\text{downside}}[P(q)] &= E[DD[P(q)]], \end{aligned} \tag{2.3}$$

where $DD[P(q)]$ is the downside deviation of profit from the expected profit which is defined as follows: $DD[P(q)] = E[(E[P(q)] - P(q))^+]$. After simplification, we have the following:

$$S_{\text{downside}}[P(q)] = p\int_0^{q-\int_0^q F(x)\mathrm{d}x}\left[q - \int_0^q F(x)\mathrm{d}x - x\right]\mathrm{d}F(x)$$
$$= p\int_0^{\lambda(q)}[\lambda(q) - x]\mathrm{d}F(x). \tag{2.4}$$

Moreover, notice that $\lambda(q) = q - \int_0^q F(x)\mathrm{d}x = q - qF(q) + \int_0^q xf(x)\mathrm{d}x$. As a result, we can rewrite $\lambda(q)$ below:

$$\lambda(q) = q[1 - F(q)] + \int_0^q xf(x)\mathrm{d}x.$$

Observe that $\lim_{q\to\infty} q[1 - F(q)] = 0$ because the demand has finite moments which means $\int_0^\infty xf(x)\mathrm{d}x<\infty$ and $\lim_{q\to\infty}\int_q^\infty xf(x)\mathrm{d}x = 0$. Since $0 \le q[1-F(q)] \le \int_q^\infty xf(x)\mathrm{d}x$, we have $\lim_{q\to\infty} q[1-F(q)]=0$. In short, we can further express $\lim_{q\to\infty}\lambda(q)$ as follows:

$\lim_{q\to\infty}\lambda(q) = 0 + \int_0^\infty xf(x)\mathrm{d}x = e$, where e is the expected demand.[1]

As a consequence, taking limit on $S_{\text{downside}}[P(q)]$ in (2.4) yields:

$$\lim_{q\to\infty} S_{\text{downside}}[P(q)] = p\left(\int_0^e [e - x]\mathrm{d}F(x)\right).$$

Since

$$\int_0^e [e - x]\mathrm{d}F(x) = \int_0^\infty [e - x]\mathrm{d}F(x) - \int_e^\infty [e - x]\mathrm{d}F(x)$$
$$= e - e - \int_e^\infty [e - x]\mathrm{d}F(x).$$

Define $L(e) = \int_e^\infty (x - e)\mathrm{d}F(x)$ which represents the right linear loss function. We have the following:

$$\lim_{q\to\infty} S_{\text{downside}}[P(q)] = pL(e).$$

Similarly, following the approach in Chen and Federgruen (2000), we explore the structural property of the variance of profit by revising it below:

$$V[P(q)] = p^2\left(q^2[1 - F(q)] + \int_0^q x^2f(x)\mathrm{d}x - \left[\int_0^q [1 - F(x)]\mathrm{d}x\right]^2\right).$$

[1] To avoid confusion, we use exp( ) to represent exponent in this chapter.

Employing integration by parts, we can further revise $V[P(q)]$in the following:

$$V[P(q)] = p^2\left(q^2[1-F(q)] + \int_0^q x^2 f(x)\mathrm{d}x - \left[[1+F(x)]x|_0^q + \int_0^q xf(x)\mathrm{d}x\right]^2\right).$$

Since we consider the demand has finite moments, we have the following:

$$\sum_{\lim\ q\to\infty} V[P(q)] = p^2\left(0 + E[x^2] - \left[[1+F(x)]x|_0^\infty + \int_0^\infty xf(x)\mathrm{d}x\right]^2\right)$$
$$= p^2\left(E[x^2] - E[x]^2\right).$$

Thus,

$$\sum_{\lim\ q\to\infty} V[P(q)] = p^2 v.$$

Proposition 2.2 summarizes the core structural properties of $E[P(q)]$, $S_{\text{downside}}[P(q)]$ and $V[P(q)]$.

Proposition 2.2[2] *(a)* $E[P(q)]$ *is a concave function of* q. *(b)* $S_{\text{downside}}[P(q)]$ *is an increasing function of* q, *and it is bounded above by* $pL(e)$. *(c)* $V[P(q)]$ *is an increasing function of* q, *and it is bounded above by* $p^2 v$.

Proof of Proposition 2.2 (a) Checking the second order derivative with respect to q shows that $\mathrm{d}^2E[P(q)]/\mathrm{d}q^2 = -pf(q) < 0$. (b) and (c) The results come from checking the respective first order derivatives with respect to q and also follow the checking of the limits when the order quantity q approaches infinity.

Since $E[P(q)]$ is a concave function, we can derive the closed-form expression for the expected profit-maximizing order quantity by solving the first order condition of $\mathrm{d}E[P(q)]/\mathrm{d}q = 0$:

$$q_E^* = F^{-1}[(p-c)/p]. \tag{2.5}$$

As an example, we consider the case when demand follows a normal distribution with mean μ and variance σ^2. Define:

$$\alpha(q) = (q-\mu)/\sigma, \tag{2.6}$$

[2] Notice that the concavity of expected profit function is a standard result in the literature. The result on variance of profit is first proven by Chen and Federgruen (2000). Notice that when we incorporate the loss of goodwill opportunity cost of stockout, the variance of profit is no longer monotone (see Choi et al. 2008; Wu et al. 2009).

$\phi(\cdot)$ is the standard normal density function, $\Phi(\cdot)$ is the standard normal cumulative distribution function, $\Psi(y) = \int_y^{\infty} (x - y)\mathrm{d}\Phi(x)$ is the standard normal right linear loss function,

$$\overline{\omega}(q) = q - \int_{-\infty}^{q} x\,\mathrm{d}\Phi(x). \tag{2.7}$$

We further define the following:

$$\xi[\alpha(q)] = \alpha(q)\phi[\alpha(q)] + [1 + \alpha(q)^2]\Phi[\alpha(q)] - (\phi[\alpha(q)] + \alpha(q)\Phi[\alpha(q)])^2], \tag{2.8}$$

$$\varsigma[q] = \left\{ \left[q - \int_{-\infty}^{q} x\,\mathrm{d}\Phi(x) - \mu \right] \Phi(\overline{\omega}(q)) + \sqrt{\frac{\sigma^2}{2\pi}} \exp\left(\frac{-(\overline{\omega}(q) - \mu)^2}{2\sigma^2} \right) \right\}. \tag{2.9}$$

Under the normally distributed demand, we can easily derive the expected profit, the variance of profit, and the semideviation of profit as follows:

$$E[P_{\text{normal}}(q)] = p\mu - cq - p\sigma\Psi[\alpha(q)], \tag{2.10}$$

$$V[P_{\text{normal}}(q)] = p^2\sigma^2\xi[\alpha(q)], \tag{2.11}$$

$$S_{\text{downside}}[P_{\text{normal}}(q)] = p\varsigma(q). \tag{2.12}$$

Solving $\mathrm{d}E[P_{\text{normal}}(q)]/\mathrm{d}q = 0$ gives the expected profit-maximizing quantity for the case when demand is normally distributed:

$$q^*_{E,\text{normal}} = \mu - \sigma\Phi^{-1}[(p - c)/p]. \tag{2.13}$$

2.2 Mean-Risk Objective Functions

In this section, we study the newsvendor problem under a mean-risk objective. For the mean-risk objective, we would consider both the semideviation of profit and the variance of profit as the measures for risk, i.e., we consider both the mean-semideviation (MS) and mean-variance (MV) models. Before we proceed to examine each case, we have Definition 2.1.

Definition 2.1 The efficient region under mean-risk framework for risk-averse retailer is denoted by Ω_j, $j = \{\text{MV}, \text{MS}\}$ with which for any $(q_1, q_2) \in \Omega_j$ and $q_1 \neq q_2$, if $E[P(q_1)] \geq E[P(q_2)]$, then $R_j(q_1) \geq R_j(q_2)$, where $R_{\text{MV}}(\cdot) = V[p(\cdot)]$, and $R_{\text{MS}}(\cdot) = S_{\text{downside}}[p(\cdot)]$.

2.2.1 The MS Model

We first consider the case when the semideviation of profit is taken as the risk measure. We define the mean-semideviation utility function for the risk-averse retailer as $U_{\mathrm{MS}}(E[P(q)], S_{\mathrm{downside}}[P(q)])$, where $U_{\mathrm{MS}}(E[P(q)], S_{\mathrm{downside}}[P(q)])$ is increasing in $E[P(q)]$ and decreasing in $S_{\mathrm{downside}}[P(q)]$. Define q^*_{MS} as the optimal order quantity which maximizes $U_{\mathrm{MS}}(E[P(q)], S_{\mathrm{downside}}[P(q)])$. We can derive Proposition 2.3.

Proposition 2.3 *(a) The efficient region* $\Omega_{\mathrm{MS}} = [0, q^*_E]$. *(b) If* $U_{\mathrm{MS}}(E[P(q)], S_{\mathrm{downside}}[P(q)])$ *is a concave function,* q^*_{MS} *uniquely exists within* Ω_{MS}.

Proof of Proposition 2.3 (a) Observe from the analytical results in Sect. 2.1 that the region with which both $E[P(q)]$ and $S_{\mathrm{downside}}[P(q)]$ increase $\forall q$ is $[0, q^*_E]$, we thus have $\Omega_{\mathrm{MS}} = [0, q^*_E]$. (b) Notice that

$$\frac{\mathrm{d}U_{\mathrm{MS}}}{\mathrm{d}q} = \frac{\partial U_{\mathrm{MS}}}{\partial E[P(q)]}\frac{\partial E[P(q)]}{\partial q} + \frac{\partial U_{\mathrm{MS}}}{\partial S_{\mathrm{downside}}[P(q)]}\frac{\partial S_{\mathrm{downside}}[P(q)]}{\partial q}.$$

Simple derivation yields:

$$\frac{\mathrm{d}U_{\mathrm{MS}}}{\mathrm{d}q} = \frac{\partial U_{\mathrm{MS}}}{\partial E[P(q)]}[(p-c) - pF(q)] + \frac{\partial U_{\mathrm{MS}}}{\partial S_{\mathrm{downside}}[P(q)]}[p(1 - F(q)F[\lambda(q)]].$$

When $q \to 0^+$, we have the following: $\left.\frac{\mathrm{d}U_{\mathrm{MS}}}{\mathrm{d}q}\right|_{q\to 0^+} = \frac{\partial U}{\partial E[P(q)]}(p-c) > 0$ because $\frac{\partial U}{\partial E[P(q)]} > 0$ and $(p-c) > 0$. When $q = q^*_E$, we have the following: $\left.\frac{\mathrm{d}U_{\mathrm{MS}}}{\mathrm{d}q}\right|_{q=q^*_E} = \frac{\partial U_{\mathrm{MS}}}{\partial S_{\mathrm{downside}}[P(q)]}(p[1 - F(q^*_E)]F[\lambda(q^*_E)]) \le 0$ because $\frac{\partial U_{\mathrm{MS}}}{\partial S_{\mathrm{downside}}[P(q)]} < 0$ and $p[1 - F(q^*_E)]$ $F[\lambda(q^*_E)] \ge 0$. As a result, if $U_{\mathrm{MS}}(E[P(q)], S_{\mathrm{downside}}[P(q)])$ is concave, q^*_{MS} must exist uniquely between 0 and q^*_E, i.e., within Ω_{MS}.

Examples of the MS objective function for risk-averse retailer are shown in Examples 2.1 and 2.2 below.

Example 2.1 (Linear mean-semideviation objective function): $U_{\mathrm{MS}}(E[P(q)], S_{\mathrm{downside}}[P(q)]) = E[P(q)] - k_{\mathrm{MS}}S_{\mathrm{downside}}[P(q)]$, where k_{MS} is a positive constant which reflects the degree of risk aversion of the retailer. To be specific, a larger k_{MS} implies a more risk-averse retailer.

Example 2.2 (Multiplicative mean-semideviation objective function): $U_{\mathrm{MS}}(E[P(q)], S_{\mathrm{downside}}[P(q)]) = E[P(q)]S_{\mathrm{downside}}[P(q)]^{-n_{\mathrm{MS}}}$, where n_{MS} is a positive constant. A larger n_{MS} means a more risk-averse retailer.

2.2.2 The MV Model

We now consider the case when the variance of profit is taken as the risk measure. We define the mean-variance utility function for the risk-averse retailer as $U_{\text{MV}}\,(E[P(q)],V[P(q)])$, where $U_{\text{MV}}(E[P(q)],V[P(q)])$ is increasing in $E[P(q)]$ and decreasing in $V[P(q)]$. Similar to the MS framework, we can derive Proposition 2.4.

Proposition 2.4 *(a) The efficient region* $\Omega_{\text{MV}}=[0,q_E^*]$. *(b) If* $U_{\text{MV}}(E[P(q)],V[P(q)])$ *is a concave function,* q_{MV}^* *uniquely exists within* Ω_{MV}.

Proof of Proposition 2.4 Similar to the approach adopted to prove Proposition 2.3.

Similar to the MS framework, examples of the MV objective function for risk-averse retailer are shown in Examples 2.3 and 2.4 below (notice that for the variance of profit, we have taken the square root to make it in the same order as the expected profit which enhances implementation and assessment).

Example 2.3 (Linear mean-variance objective function)[3]: $U_{\text{MV}}(E[P(q)],V[P(q)]) = E[P(q)] - k_{\text{MV}}\sqrt{V[P(q)]}$, where k_{MV} is a positive constant which reflects the degree of risk aversion of the retailer.

Example 2.4 (Multiplicative mean-variance objective function): $U_{\text{MV}}(E[P(q)],$ $V[P(q)]) = E[P(q)]\sqrt{V[P(q)]}^{-n_{\text{MV}}}$, where n_{MV} is a positive constant which reflects the degree of risk aversion of the retailer.

As a remark, if we include the loss of goodwill opportunity cost of stockout into the model, the efficient region will no longer be $[0,q_E^*]$. This result is brought by the change of the structural property of the variance of profit function which is no longer monotone increasing (even though the expected profit function remains concave). In particular, both Choi et al. (2008) and Wu et al. (2009) show that the optimal ordering quantity under the mean-variance model can exceed q_E^* in the presence of the loss of goodwill opportunity cost of stockout.

2.3 Efficient Frontiers

We investigate the construction of the efficient frontiers under the MS and MV models in this section. An efficient frontier gives the plot of $E[P(q)]$ versus the risk measure and shows the trade-off between $E[P(q)]$ and risk in the efficient region. Every point on the efficient frontier gives us a non-inferior efficient solution point. For example, under the MS model, for a risk-averse retailer, any two order quantities q_1 and q_2 on the efficient frontier should possess the following properties: If $E[P(q_1)]\geq E[P(q_2)]$, then $S_{\text{downside}}[P(q_1)]\geq S_{\text{downside}}[P(q_2)]$ and vice versa. For the risk-averse retailer, as we have proven earlier in this chapter, the efficient region is $[0,q_E^*]$. As a result, for any risk-averse retailer, an optimal decision should be selected from the set of efficient solutions on the efficient frontier within $[0,q_E^*]$. In order to construct the efficient frontier, we can follow the approach by Choi et al. (2008) and define the slope of a point on the efficient frontier as follows:

[3] This utility function is similar to the ones employed by Lau (1980) and Choi et al. (2008).

$$\text{For the MS Model}: \gamma_{\text{MS}} = \frac{\mathrm{d}E[P(q)]/q}{\mathrm{d}S_{\text{downside}}[P(q)]/q}. \tag{2.14}$$

$$\text{For the MV Model}: \gamma_{\text{MV}} = \frac{\mathrm{d}E[P(q)]/q}{\mathrm{d}V[P(q)]/q}. \tag{2.15}$$

We have Proposition 2.5.

Proposition 2.5

$$(a)\ \gamma_{\text{MS}} = \frac{(p-c) - pF(q)}{pF(-q)F[\lambda(q)]}, \qquad (b)\ \gamma_{\text{MV}} = \frac{(p-c) - pF(q)}{2p^2F(-q)n(q)}. \tag{2.16}$$

With Proposition 2.5, we can numerically generate the efficient frontiers for the MS and MV models, respectively. Observe that when $q = 0$, we have $\gamma_{\text{MS}} \to \infty$ and $\gamma_{\text{MV}} \to \infty$. As a result, $q = 0$ is called an improper solution for both the MS and MV models.

We now proceed to derive the analytical expression of the efficient frontier for the case when the demand is normally distributed. Under the MS model, from (2.12), we have $S_{\text{downside}}[P_{\text{normal}}(q)] = p\varsigma(q)$. Define: $\varsigma^{-1}(\cdot)$ be the inverse function of $\varsigma(\cdot)$.

Simple manipulation yields (2.17):

$$\frac{S_{\text{downside}}[P_{\text{normal}}(q)]}{p} = \varsigma(q) \Leftrightarrow q = \varsigma^{-1}\left(\frac{S_{\text{downside}}[P_{\text{normal}}(q)]}{p}\right). \tag{2.17}$$

Putting (2.17) into (2.10) yields the efficient frontier under the MS model

$$E[P_{\text{normal}}(q)] = p\left(\mu - \sigma\Psi\left[\alpha\left[\varsigma^{-1}\left(\frac{S_{\text{downside}}[P_{\text{normal}}(q)]}{p}\right)\right]\right]\right) - c\varsigma^{-1}\left(\frac{S_{\text{downside}}[P_{\text{normal}}(q)]}{p}\right), \tag{2.18}$$

and the efficient region Ω_{MS} is $[0, q^*_{E,\text{normal}}]$.

Similar, for the MV model, we define:

$$\tau(q) = \xi[\alpha(q)], \tag{2.19}$$

$$\tau^{-1}(\cdot) \text{ be the inverse function of } \tau(\cdot). \tag{2.20}$$

From (2.11), we have $V[P_{\text{normal}}(q)] = p^2\sigma^2\xi[\alpha(q)]$, it is easy to show that:

$$q = \tau^{-1}\left(\frac{V[P_{\text{normal}}(q)]}{p^2\sigma^2}\right) \tag{2.21}$$

Putting (2.21) into (2.10) yields the efficient frontier under the MV model

$$E[P_{\text{normal}}(q)] = p\left(\mu - \sigma\Psi\left[\alpha\left[\tau^{-1}\left(\frac{V[P_{\text{normal}}(q)]}{p^2\sigma^2}\right)\right]\right]\right) - c\tau^{-1}\left(\frac{V[P_{\text{normal}}(q)]}{p^2\sigma^2}\right), \tag{2.22}$$

and the efficient region Ω_{MV} is $[0, q^*_{E,\text{normal}}]$.

2.4 Mean-Semideviation and Mean-Variance Models[4]

From the above discussions, we notice that the mean-semideviation model and the mean-variance model look very similar and the efficient regions under both models are the same. In the following, we show that when the two models are "fairly" compared, they in fact would have the same solution. First of all, we define the following MS and MV models:

$$(P_{\text{MS}}) \quad \min_q S_{\text{downside}}[P(q)] \quad \text{subject to } E[P(q)] \geq \eta,$$
$$(P_{\text{MV}}) \quad \min_q V[P(q)] \quad \text{subject to } E[P(q)] \geq \eta,$$

where η is the minimum expected profit target that the retailer would like to achieve.

To have meaningful models, η must be bounded above by the maximum achievable expected profit, i.e.,

$$\eta \leq E[P(q^*_E)]. \tag{2.23}$$

Define:

$$q_{\eta*} = \arg_{q \leq q^*_E} \{E[P(q)] = \eta\}. \tag{2.24}$$

We denote q_{MS^*} and q_{MV^*} as the optimal ordering quantities for (P_{MS}) and (P_{MV}), respectively. By noting (1) the concavity of the expected profit function and (2) the monotone increasing property of both the variance of profit and the semideviation of profit, we have Proposition 2.6.

Proposition 2.6

$$q_{\text{MS}^*} = q_{\text{MV}^*} = q_{\eta^*}.$$

Proposition 2.6 is an interesting result which reveals that despite the potential theoretical flaw associated with the variance measure, the optimal stocking quantity

[4] This part comes mainly from Choi and Chiu (2012) and interested reader is referred to it for more discussions. We acknowledge Elsevier for granting us the authorship right to reuse these materials in this book format.

for the MV model is the same as that of the MS case. As a result, in the subsequent chapters, for the sake of brevity, we will just employ one of these measures to conduct analysis.

As a remark, if we consider the MS and MV problems as the following alternative settings and define q_{MS,A^*} and q_{MV,A^*} as the optimal solutions for $(P_{MS,Alternative})$ and $(P_{MV,Alternative})$, respectively, it is easy to prove that $q_{MV,A^*} \leq q_{MS,A^*} \leq q_E^*$.

$$(P_{MS,Alternative}) \quad \max_q E[P(q) \quad \text{subject to } S[P(q)] \leq k,$$

$$(P_{MV,Alternative}) \quad \max_q E[P(q)] \quad \text{subject to } \sqrt{V[P(q)]} \leq k,$$

where $k \geq 0$ is a risk tolerance threshold which represents the maximum amount of risk that the retailer is willing to take. We set $\sqrt{V[P(q)]} \leq k$ as the constraint on risk to make it "closer" to the constraint $S[P(q)] \leq k$ (to avoid the squaring effect of the variance). However, even with this change, we argue that $(P_{MS,Alternative})$ and $(P_{MV,Alternative})$ actually should not be compared because a risk tolerance level under $(P_{MV,Alternative})$ is a threshold to compare with the standard deviation of profit, whereas a risk tolerance level under $(P_{MS,Alternative})$ is a threshold to compare with the semideviation of profit. In general, decision makers should understand this point when they decide the right value of k for the respective problem. Thus, in order to have a fair treatment, the value of "k" for $(P_{MS,Alternative})$ and $(P_{MV,Alternative})$ should not be the same.

2.5 Numerical Analyses

To supplement the analytical findings derived above, we conduct several numerical analyses. We first consider the example with the following parameters: Demand follows a normal distribution with mean $\mu = 100$, standard deviation $\sigma = 20$. The unit retail selling price $p = 10$, the unit product cost $c = 3$. With these parameters, the expected profit-maximizing order quantity can be calculated by (2.13) as follows,

$$q_{E,normal}^* = 100 - 20\Phi^{-1}[(10-3)/10] = 110.$$

To understand how the expected profit and the level of risk vary with order quantity q, we present Table 2.1. As a remark, since the analysis results for the MS and MV models are similar, we just present the numerical analysis using the MV model in which the level of risk is measured by the variance of profit (or equivalently the standard deviation of profit).

From Table 2.1, we can see that the efficient region is bounded between 0 and 110. For any order quantity q within this efficient region, the corresponding expected profit and level of risk is non-inferior to one another because the order

Table 2.1 The expected profit $E[P(q)]$, the variance of profit $V[P(q)]$, and the standard deviation of profit $sd[P(q)]$ with different order quantity q

q	$E[P(q)]$	$V[P(q)]$	$sd[P(q)]$
0	0.00	0.00	0.03
10	70.00	0.01	0.10
20	140.00	0.12	0.35
30	209.99	1.12	1.06
40	279.92	8.13	2.85
50	349.60	47.81	6.91
60	418.30	227.87	15.10
70	484.14	879.52	29.66
80	543.34	2,735.93	52.31
90	590.44	6,820.63	82.59
100	620.21	13,633.80	116.76
$110 = q_E^*$	630.44	22,137.63	148.79
120	623.34	30,043.51	173.33
130	604.14	35,534.95	188.51
140	578.30	38,407.85	195.98
150	549.60	39,551.04	198.87
160	519.92	39,900.14	199.75
170	489.99	39,982.51	199.96
180	460.00	39,997.59	199.99
190	430.00	39,999.74	200.00
200	400.00	39,999.98	200.00
210	370.00	40,000.00	200.00
220	340.00	40,000.00	200.00
230	310.00	40,000.00	200.00
240	280.00	40,000.00	200.00
250	250.00	40,000.00	200.00
260	220.00	40,000.00	200.00
270	190.00	40,000.00	200.00
280	160.00	40,000.00	200.00
290	130.00	40,000.00	200.00
300	100.00	40,000.00	200.00
310	70.00	40,000.00	200.00
320	40.00	40,000.00	200.00
330	10.00	40,000.00	200.00
340	−20.00	40,000.00	200.00
350	−50.00	40,000.00	200.00
360	−80.00	40,000.00	200.00
370	−110.00	40,000.00	200.00
380	−140.00	40,000.00	200.00
390	−170.00	40,000.00	200.00
400	−200.00	40,000.00	200.00

quantity which leads to a higher expected profit is also associated with a higher level of risk. Figure 2.1 shows the expected profit E[P(q)], and the standard deviation of profit $sd[P(q)]$ plotted against the order quantity q.

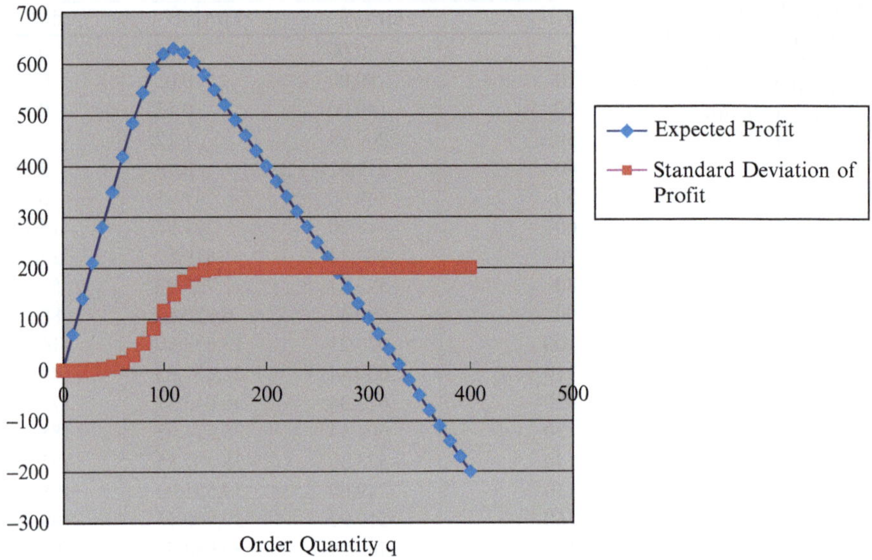

Fig. 2.1 The expected profit $E[P(q)]$ and the standard deviation of profit $\mathrm{sd}[P(q)]$ plotted against the order quantity q

From Fig. 2.1, we can see that the expected profit is a concave function and the standard deviation of profit is an increasing function. It reaches a steady state when the order quantity becomes rather large (around 180). In order to show how the expected profit and standard deviation of profit change with reference to the benchmarking case of expected profit maximization, we define the following for any q:

$$\Delta E[P(q)] = E[P(q)] - E[P(q_E^*)],$$

$$\Delta \mathrm{sd}[P(q)] = \mathrm{sd}[P(q)] - \mathrm{sd}[P(q_E^*)].$$

To get a clearer picture, we also define the percentage changes as follows,

$$\%\Delta E[P(q)] = \frac{E[P(q)] - E[P(q_E^*)]}{E[P(q_E^*)]} \times 100\%,$$

$$\%\Delta \mathrm{sd}[P(q)] = \frac{\mathrm{sd}[P(q)] - \mathrm{sd}[P(q_E^*)]}{\mathrm{sd}[P(q_E^*)]} \times 100\%.$$

Table 2.2 shows the numerical results on $\Delta E[P(q)]$, $\Delta \mathrm{sd}[P(q)]$, $\%\Delta E[P(q)]$, $\%\Delta \mathrm{sd}[P(q)]$ with different ordering quantity q. Figures 2.2 and 2.3 further plot in curves the changes $\Delta E[P(q)]$, $\Delta \mathrm{sd}[P(q)]$ and percentage changes $\%\Delta E[P(q)]$, $\%\Delta \mathrm{sd}[P(q)]$, respectively.

Table 2.2 $\Delta E[P(q)]$, $\%\Delta E[P(q)]$, $\Delta \text{sd}[P(q)]$ and $\%\Delta \text{sd}[P(q)]$ for q within the efficient region

q	$\Delta E[P(q)]$	$\%\Delta E[P(q)]$	$\Delta \text{sd}[P(q)]$	$\%\Delta \text{sd}[P(q)]$
0	−630.44	−100.00	−148.76	−99.98
10	−560.44	−88.90	−148.69	−99.93
20	−490.44	−77.79	−148.44	−99.76
30	−420.45	−66.69	−147.73	−99.29
40	−350.52	−55.60	−145.94	−98.08
50	−280.84	−44.55	−141.88	−95.36
60	−212.14	−33.65	−133.69	−89.85
70	−146.3	−23.21	−119.13	−80.07
80	−87.1	−13.82	−96.48	−64.84
90	−40	−6.34	−66.2	−44.49
100	−10.23	−1.62	−32.03	−21.53
110	0	0.00	0	0.00

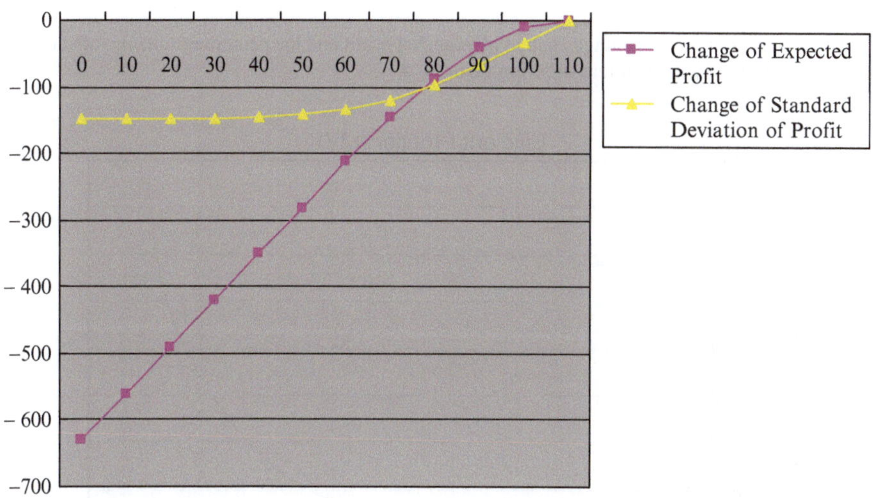

Fig. 2.2 A plot of the change of expected profit and the change of standard deviation of profit with different ordering quantity q

From Table 2.2, one interesting result is that: By ordering at a quantity q less than the expected profit-maximizing order quantity, the retailer has a good opportunity of reducing the level of risk. In fact, for many cases, the reduction of risk is much more than the decrease of expected profit. For instance, from Fig. 2.2, we can see that for ordering quantity equals 80, 90, and 100, the reduction of standard deviation of profit is larger than the reduction of expected profit. One can hence argue that by ordering at these quantities, the reduction of risk is more significant compared to the drop in expected profit. If we adopt the percentage change as a way to compare the significance of changes, Fig. 2.3 shows another

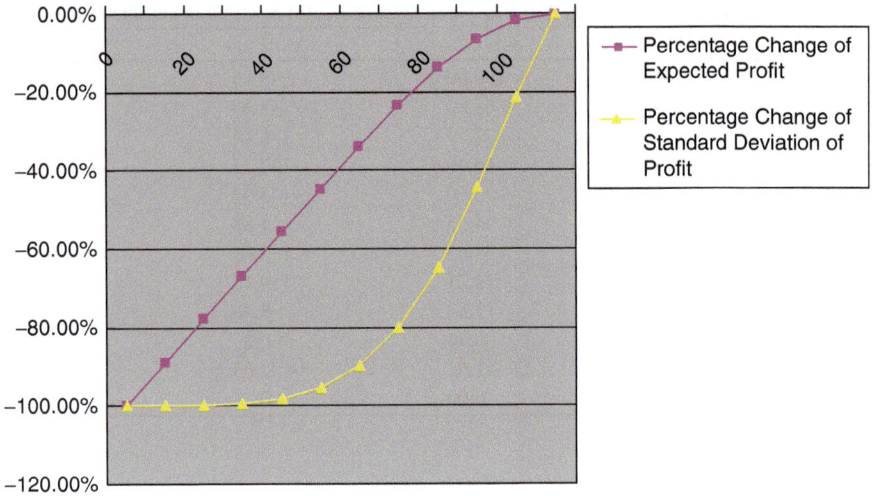

Fig. 2.3 A plot of the percentage change of expected profit and the percentage change of standard deviation of profit with different ordering quantity q

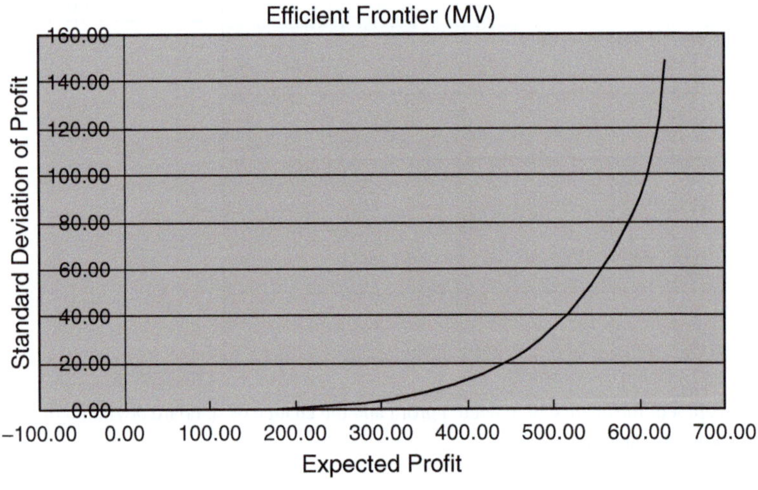

Fig. 2.4 The efficient frontier under mean-variance model

very interesting result in which the percentage reduction of risk (i.e., standard deviation of profit) is always larger than or equal to that of the decrease of expected profit. For some cases, the percentage reduction of risk is even more significant compared to the sacrifice of expected profit. For instance, in Table 2.2, when the order quantity is 90, the percentage reduction of standard deviation of profit is 44.49% while the percentage decrease of expected profit is just 6.34%. A similar pattern can be observed for many other order quantities such as 70 and 80.

As such, it makes a lot of sense for the retailer to seriously consider ordering a quantity (within the efficient region) which differs from the expected profit-maximizing quantity (should actually be lower) because it can significantly reduce the level of risk with a relatively small decrease in the expected profit.

With the data from Table 2.2, we can further extract the data points on expected profit and standard deviation of profit out and plot in Fig. 2.4 the efficient frontier for the ordering quantity within the efficient region. This efficient frontier helps the retailer determine the best point to take by making a trade-off between the expected profit and the corresponding level of risk.

To show how demand variance and profit margin affect the mean-variance efficient frontier, we conduct sensitivity analysis. We use the same default set of data as in Section 2.5.1. We vary and set $\sigma = 5, 15, 25, 35$, and $p = 4, 8, 12, 16, 20$, respectively for exploring the impacts of σ and p. The numerical values of the resulting expected profit and standard deviation of profit are shown in Tables 2.3 and 2.4, respectively. We further plot the corresponding efficient frontiers for the two cases in Figs. 2.5 and 2.6.

From Fig. 2.5, we can observe that if the standard deviation of demand σ increases, the mean-variance efficient frontier will shift towards the left top corner. It means that the standard deviation of profit becomes bigger when σ increases. Moreover, the efficient frontier for the case with a lower σ dominates the case

Table 2.3 Expected profit and standard deviation of profit with differentσ for q within the efficient region

σ	Expected profit	Standard deviation of profit
5	0.00	0.00
	70.00	0.00
	140.00	0.00
	210.00	0.00
	280.00	0.00
	350.00	0.00
	420.00	0.00
	490.00	0.00
	560.00	0.09
	629.58	3.77
	680.05	29.19
15	0.00	0.00
	70.00	0.00
	140.00	0.01
	210.00	0.05
	280.00	0.26
	349.98	1.12
	419.82	3.91
	488.73	11.32
	553.64	27.20
	607.33	53.86
	640.16	87.57
	647.33	118.48

(continued)

Table 2.3 (continued)

σ	Expected profit	Standard deviation of profit
25	0.00	0.44
	69.99	1.07
	139.95	2.43
	209.81	5.14
	279.32	10.19
	347.88	18.87
	414.19	32.66
	475.97	52.80
	529.95	79.55
	572.39	111.61
	600.26	145.95
	612.39	178.56
35	0.00	6.49703
	69.44	10.7313
	138.66	17.1276
	207.03	26.4166
	273.82	39.3665
	337.95	56.6588
	397.95	78.7128
	452	105.488
	498.17	136.325
	534.71	169.898
	560.37	204.337
	574.71	237.52
	578.17	267.47

Table 2.4 Expected profit and standard deviation of profit with different p for q within the efficient region

p	Expected profit	Standard deviation of profit
4	0.00	0.01
	10.00	0.04
	20.00	0.14
	30.00	0.42
	39.97	1.14
	49.84	2.77
	59.32	6.04
	67.66	11.86
	73.33	20.92
	74.18	33.03
8	0.00	0.02
	50.00	0.08
	100.00	0.28
	149.99	0.85

(continued)

Table 2.4 (continued)

p	Expected profit	Standard deviation of profit
	199.94	2.28
	249.68	5.53
	298.64	12.08
	345.31	23.73
	386.67	41.84
	418.35	66.07
	436.17	93.41
	438.35	119.03
12	0.00	0.03
	90.00	0.13
	180.00	0.42
	269.99	1.27
	359.91	3.42
	449.52	8.30
	537.96	18.11
	622.97	35.59
	700.00	62.77
	762.53	99.10
	804.25	140.12
	822.53	178.54
16	0.00	0.04
	130.00	0.17
	260.00	0.56
	389.98	1.69
	519.88	4.56
	649.36	11.06
	777.28	24.15
	900.62	47.45
	1,013.34	83.69
	1,106.71	132.14
	1,172.34	186.82
	1,206.71	238.06
	1,213.34	277.33
20	0.00	0.06
	170.00	0.21
	340.00	0.70
	509.98	2.11
	679.85	5.70
	849.20	13.83
	1,016.60	30.19
	1,178.28	59.31
	1,326.67	104.61
	1,450.88	165.17
	1,540.42	233.53
	1,590.88	297.57
	1,606.67	346.66

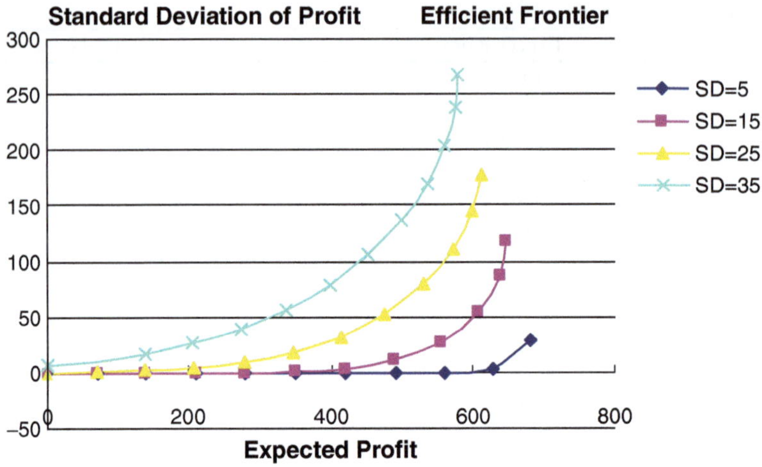

Fig. 2.5 The efficient frontiers under mean-variance model with different standard deviation σ (SD)

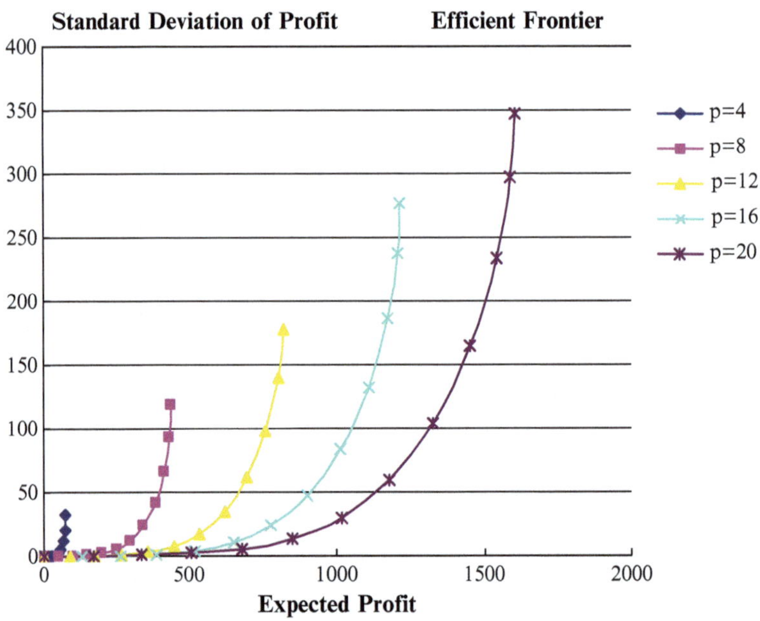

Fig. 2.6 The efficient frontiers under mean-variance model with different p

with a larger σ in the mean-variance domain. In other words, a lower σ is always more desirable for the decision maker with the consideration of mean-variance efficient frontier. Another pattern is observed in Fig. 2.6 for the impact brought by different retail selling price p on the efficient frontier. To be specific, we can see that if p becomes bigger, the efficient frontier will move to the right. It is also interesting to note that the efficient frontier for the case with a bigger p has a higher expected profit but also a higher variance of profit compared to the one with a smaller p. We hence cannot conclude that the case with a higher p is always better off than the case with a small p.

2.6 Conclusion and Remarks

We have explored in this chapter the single-period inventory problem under the mean-risk framework. We explicitly study both the MS and MV models. From our analysis, we can see that the incorporation of risk measure is important not only because it can capture the stochastic nature of the problem, but it can also include the retailer's risk attitude into the decision-making process. A tailor-fit optimal decision for each specific decision maker can hence be made. We derive the efficient region for both the MS and MV models and also illustrate how to construct an efficient frontier for both cases. By considering the normally distributed demand, we analytically derive the corresponding efficient frontier and also numerically conduct the mean-risk analysis. Our findings illustrate that it makes sense for the retailer to consider ordering at a quantity (within the efficient region) which is lower than the expected profit-maximizing quantity because this action can significantly reduce the level of risk with a relatively small sacrifice on expected profit. We have also shown the equivalence of the solutions between the mean-variance and mean-downside-risk newsvendor problems. As such, even though the mean-variance approach suffers the theoretical flaw of counting both the upside and downside deviations, its nice structural properties allow itself to have the same solution as compared to the mean-downside-risk model's for the properly formulated newsvendor problem. In the subsequent chapters, we mainly focus on mean-variance analysis.

References

Chen, F., & Federgruen, A. (2000). *Mean-variance analysis of basic inventory models.* Working paper, Columbia University.

Choi, T. M., & Chiu, C. H. (2012). Mean-downside-risk and mean-variance newsvendor models: implications for sustainable fashion retailing. *International Journal of Production Economics, 135*, 552–560.

Choi, T. M., Li, D., & Yan, H. (2008). Mean-variance analysis for the newsvendor problem. *IEEE Transactions on Systems, Man, and Cybernetics, Part A - Systems and Humans, 38*, 1169–1180.

Lau, H. S. (1980). The newsboy problem under alternative optimization objectives. *Journal of the Operational Research Society, 31*, 525–535.

Wu, J., Li, J., Wang, S., & Cheng, T. C. E. (2009). Mean–variance analysis of the newsvendor model with stockout cost. *Omega, 37*, 724–730.

Chapter 3
Mean-Risk Analysis of Multiperiod Inventory Problems

In this chapter, we carry out mean-risk analysis of multiperiod inventory problems. We select the well-known (R, nQ) multiperiod inventory replenishment model (see Chen and Zheng 1994, 1998; Larsen and Kiesmüller 2007; Li and Sridharan 2008; Shang and Zhou 2010; Lagodimos et al. 2012 and the references therein for more details of the recent developments and extensions of this model) as an example to demonstrate how to perform a mean-risk analysis for multiperiod inventory problems. As shown later on in this chapter, the mean-risk analysis of multiperiod inventory problems is very different from the mean-risk analysis of single-period analysis, in terms of problem formulations and methodology applied. In particular, the (R, nQ) model considers an infinite-horizon replenishment problem under which the total profit/cost is infinite, too. Therefore, the expected (total) profit and the variance of (total) profit cannot be used directly as the "mean" and the "risk," respectively, in the mean-risk analysis of the (R, nQ) model. To perform the mean-risk analysis, we take the long-run average profit as the "mean," and propose the variance of on-hand inventory and the variance of one-period profit as "risk" of the (R, nQ) model. We first derive the closed form expressions of the long-run average profit, the variance of on-hand inventory, and the variance of one-period profit. Then, we apply the numerical analysis to demonstrate how to construct the efficient frontier, in the mean-risk sense.

3.1 The (R, nQ) Model

The (R, nQ) model considers a single-location, single-product, multiple periods inventory problem. Consider a retailer which sells a product in a very long selling season (infinite horizon) to the customers. The product is provided by an outside supplier, and the retailer reviews the inventory periodically. Demands in different review periods are independent and identically distributed (i.i.d.) random variables. The retailer adopts an (R, nQ) policy to replenish inventory. To be specific, whenever the inventory position (on-hand inventory plus outstanding orders

T.-M. Choi and C.-H. Chiu, *Risk Analysis in Stochastic Supply Chains: A Mean-Risk Approach*, International Series in Operations Research & Management Science 178,
DOI 10.1007/978-1-4614-3869-4_3, © Springer Science+Business Media New York 2012

minus backorders) is at or below a reorder point[1] r, the retailer orders a minimum integer multiple of the "per batch order size" Q to raise the inventory position to above r. The delivery lead time from the supplier to the retailer is given by l, where l is a nonnegative integer. To be specific, under the (R, nQ) policy, an order placed at the beginning of period t, where t is a nonnegative integer, arrives at the beginning of period $t + l$. An ordering cost C is incurred for each batch of Q units ordered. That is, if the order size is nQ units in a particular replenishment, then the total ordering cost for this replenishment is given by nC. At the end of each period, any on-hand inventory results in a holding cost h per unit and any backlogged demand incurs a backorder cost b per unit. The retail price of the product is p and the product cost is c per unit. The planning horizon is infinite. The decision variables are r and Q, and the retailer would like to maximize the long-run average total profit of the retailer with the consideration of some measure of risk to be specified later. As a remark, Chen and Federgruen (2000) study the cost minimization problem of the (R, nQ) model with risk consideration; we follow a similar approach to perform the mean-risk analysis of the (R, nQ) model under the profit model. Since it is a well-known fact that the cost and the profit models will lead to different optimal decisions when the objective function is neither linear nor exponential, our analysis is different from Chen and Federgruen (2000)'s.

Before we go into the detailed mean-risk analysis, we first review some commonly known results about the (R, nQ) model. Define

$I(t)$	= inventory position at the beginning of period t.
I	= steady-state inventory position.
$L(t)$	= inventory level at the end of period t ($L(t) > 0$ means that the retailer has on-hand inventory, and $L(t) < 0$ means that there is backorder).
L	= steady-state inventory level.
D	= one-period demand, a discrete random variable, with mean $E(D) = \mu$, variance $V(D) = \sigma^2$, probability mass function (pmf) $f(\cdot)$, and cumulative distribution function (cmf) $F(\cdot)$.
D_i	= demand over i periods, with pmf $f_i(\cdot)$ and cmf $F_i(\cdot)$, for $i > 1$.
$D(t, t+l)$	= total demand in periods t, $t + 1$, ..., $t + l$, which is distributed the same as D_{l+1}.
$\pi(r, Q)$	= long-run average profit of the retailer (revenue minus ordering cost, holding cost, and backorder cost).

The well-known inventory balance equation is given by

$$L(t+l) = I(t) - D(t, t+l). \tag{3.1}$$

For any given $I(t) = y$, the holding cost in period $t + l$ is $h(y - D_{l+1})^+$, and the backorder cost is $b(D_{l+1} - y)^+$, where $A^+ = \max(0, A)$. Therefore, the expected

[1] In the (R, nQ) policy, R represents the reorder point, and nQ is the order quantity in which Q is the order size per batch. In this chapter, we use r to represent the decision variable of reorder point.

inventory cost (which is equal to the sum of holding and backorder costs) in period $t + l$ for given $I(t) = y$ is

$$G(y) = E[h(y - D_{l+1})^+ + b(D_{l+1} - y)^+].$$

As the steady-state of $I(t)$ is uniformly distributed on the interval $[r + 1, r + Q]$ (see, e.g., Hadley and Whitin (1961)), the long-run average holding and backorder costs is $\sum_{y=r+1}^{r+Q} G(y)/Q$. Moreover, the long-run average ordering cost is $\mu C/Q$. Therefore, the long-run average profit is

$$\pi(r, Q) = (p - c)\mu - \sum_{y=r+1}^{r+Q} G(y)/Q - \mu C/Q.$$

It is known that $\left[\sum_{y=r+1}^{r+Q} G(y) + \mu C\right]/Q$ is jointly convex in r and Q (Zipkin 2000). Therefore, $\pi(r, Q)$ is jointly concave in r and Q.

3.2 Variance of On-Hand Inventory

We first consider the variance of on-hand inventory at the end of period as the risk measure of the retailer. Unlike the backorders and the undelivered orders, which are not yet the assets of the retailer, on-hand inventory is owned by the retailer. The retailer needs to pay its supplier in order to obtain the product, so on-hand inventory actually ties up the capital of the retailer. For some expensive products, the capital tied up in inventory can be very big. It can affect the cash flow of the retailer significantly. Therefore, if the retailer is concerned with the fluctuation in the capital tied up in on-hand inventory and the liquidity of the company, the variance of on-hand inventory is of interests. Obviously, in mean-variance perspective, a lower variance of on-hand inventory is better.

On-hand inventory at the end of period is given by L^+. We let $V_{\mathrm{I}}(r, Q)$ be the variance of L^+. Consider the following mean-risk problem of the (R, nQ) model with the long-run average profit as the "mean" and the variance of on-hand inventory as the "risk" measure:

$$\max_{r,Q} \pi(r, Q)$$
$$\text{s.t. } V_{\mathrm{I}}(r, Q) \le v, \tag{P3.1}$$

where $v > 0$ is a prespecified maximum acceptable variance of L^+ of the retailer. Besides the formulation given by (P3.1), we can also consider the following mean-risk problem of the (R, nQ) model:

$$\min_{r,Q} V_{\mathrm{I}}(r,Q)$$
$$\text{s.t. } \pi(r,Q) \geq k, \tag{P3.2}$$

where $k > 0$ is a prespecified minimum requirement of the long-run average of profit of the retailer. Consider specific $v = v_1 > 0$ and $k = k_1 > 0$, and suppose that (r_1, Q_1) is the optimal solution of (P3.1) for $v = v_1 > 0$, and $\pi(r_1, Q_1) \geq k_1 > 0$. For any $(r,Q) \neq (r_1, Q_1)$, $V_{\mathrm{I}}(r,Q) > V_{\mathrm{I}}(r_1, Q_1)$ if $\pi(r,Q) > \pi(r_1, Q_1) \geq k_1$. So (r_1, Q_1) is the optimal solution of (P3.2) for $k = k_1 > 0$. Therefore, an optimal solution of (P3.1) for a given $v > 0$ is also an optimal solution of (P3.2) for some $k > 0$, and vice versa. The pair of (v, k) is an efficient point in $\pi(r,Q) - V_{\mathrm{I}}(r,Q)$ (mean-risk) space, i.e., an efficient point of the mean-risk problem of the (R, nQ) model with the long-run average profit as the "mean" and the variance of on-hand inventory as the "risk." By varying the value of v, or k, in the above problems, we can obtain the efficient frontier. We note that v and k in the pair of (v, k) may not be a one-to-one mapping. In other words, a give $v > 0$ can pair up with different $k > 0$ to form an efficient point, and vice versa.

In the following, we focus on (P3.1) only. In steady state, the inventory balance (3.1) can be rewritten as:

$$L = I - D_{l+1}.$$

As, I is independent of D_{l+1}, we obtain:

$$E[L^+ | I = y] \equiv u_1(y) = \sum_{x=0}^{y} (y - x) f_{l+1}(x)$$

and

$$E[(L^+)^2 | I = y] \equiv u_2(y) = \sum_{x=0}^{y} (y - x)^2 f_{l+1}(x),$$

for all integer $y > 0$. For any integer $y < 0$, as $y - D_{l+1} < 0$, $L < 0$ with probability 1. Hence, $L^+ = 0$ with probability 1 and $u_1(y) = u_2(y) = 0$. Since I is uniformly distributed on the interval $[r+1, r+Q]$,

$$\begin{aligned} V_{\mathrm{I}}(r,Q) &= E[(L^+)^2] - (E[(L^+)])^2 \\ &= \frac{1}{Q}\sum_{y=r+1}^{r+Q} E[(L^+ | I = y)^2] - \left(\frac{1}{Q}\sum_{y=r+1}^{r+Q} E[L^+ | I = y]\right)^2 \\ &= \frac{1}{Q}\sum_{y=r+1}^{r+Q} u_2(y) - \left(\frac{1}{Q}\sum_{y=r+1}^{r+Q} u_1(y)\right)^2. \end{aligned} \tag{3.2}$$

Next, since $u_1(y) = yF_{l+1}(y) - \sum_{x=0}^{y} x f_{l+1}(x)$, we have:

$$\begin{aligned} u_1(y+1) - u_1(y) &= (y+1)F_{l+1}(y+1) - \sum_{x=0}^{y+1} x f_{l+1}(x) - \left[y F_{l+1}(y) + \sum_{x=0}^{y} x f_{l+1}(x) \right] \\ &= (y+1)F_{l+1}(y+1) - yF_{l+1}(y) - (y+1)f_{l+1}(x) \\ &= F_{l+1}(y). \end{aligned} \tag{3.3}$$

On the other hand, we have:

$$u_2(y+1) = \sum_{x=0}^{y+1} (y+1-x)^2 f_{l+1}(x) = \sum_{x=0}^{y} (y+1-x)^2 f_{l+1}(x),$$

so

$$\begin{aligned} u_2(y+1) - u_2(y) &= \sum_{x=0}^{y} (y+1-x)^2 f_{l+1}(x) - \sum_{x=0}^{y} (y-x)^2 f_{l+1}(x) \\ &= \sum_{x=0}^{y} [(y+1-x) + (y-x)] f_{l+1}(x) \\ &= u_1(y+1) + u_1(y). \end{aligned} \tag{3.4}$$

Moreover,

$$\begin{aligned} \sum_{y=r+2}^{r+1+Q} u_1(y) + \sum_{y=r+1}^{r+Q} u_1(y) &= \sum_{y=r+1}^{r+Q} [u_1(y+1) + u_1(y)] \\ &= \sum_{y=r+1}^{r+Q} [u_2(y+1) - u_2(y)] (\text{by}(3.4)) \\ &= u_2(r+1+Q) - u_2(r+1). \end{aligned}$$

Hence, we have:

$$\begin{aligned} \left(\frac{1}{Q} \sum_{y=r+2}^{r+1+Q} u_1(y) \right)^2 - \left(\frac{1}{Q} \sum_{y=r+1}^{r+Q} u_1(y) \right)^2 = &\left(\frac{u_2(r+1+Q) - u_2(r+1)}{Q} \right) \\ &\left(\frac{u_1(r+1+Q) - u_1(r+1)}{Q} \right). \end{aligned}$$

Therefore,

$$
\begin{aligned}
V_{\mathrm{I}}(r+1,Q)-V_{\mathrm{I}}(r,Q) &= \frac{u_2(r+1+Q)-u_2(r+1)}{Q} - \left(\frac{1}{Q}\sum_{y=r+2}^{r+1+Q} u_1(y)\right)^2 + \left(\frac{1}{Q}\sum_{y=r+1}^{r+Q} u_1(y)\right)^2 \\
&= \left(\frac{u_2(r+1+Q)-u_2(r+1)}{Q}\right)\left(1-\frac{u_1(r+1+Q)-u_1(r+1)}{Q}\right) \\
&\geq 0,
\end{aligned}
$$

where the last inequality holds because $u_2(\cdot)$ is nondecreasing (by (3.4)) and

$$
u_1(r+1+Q)-u_1(r+1) = \sum_{y=r+1}^{r+Q}[u_1(y+1)-u_1(y)] = \sum_{y=r+1}^{r+Q} F_{l+1}(y) \leq Q
$$

(by (3.3)).

Therefore, we have the following Proposition 3.1.[2]

Proposition 3.1 *For any fixed Q, $V_{\mathrm{I}}(r,Q)$ is nondecreasing in r.*

As $\pi(r,Q)$ and $V_{\mathrm{I}}(r,Q)$ are non-linear in r and Q, Problems (P3.1) and (P3.2) are non-linear programming problems. Next, for any fixed Q, let $\pi(r,Q)$ be maximized at $r = r_{\mathrm{e}}(Q)$. Define $\pi_{\mathrm{e}}(Q) = \pi(r_{\mathrm{e}}(Q),Q)$. Let

$$
r_{\mathrm{v}}(Q) = \max\{r : V_I(r,Q) \leq v\},
$$

and (r_0,Q_0) be any feasible solution to (P3.1) with $\pi_0 = \pi(r_0,Q_0)$. Define $Q_1 = \max\{Q : \pi_{\mathrm{e}}(Q) \geq \pi_0\}$. Let (r_1^*,Q^*) be the optimal solution to (P3.1).

Proposition 3.2 *(a) For any fixed Q, $r_{\mathrm{v}}(Q) \geq -Q$. (b) For any fixed Q, $r = \min\{r_{\mathrm{e}}(Q), r_{\mathrm{v}}(Q)\}$ solves (P3.1). (c) $Q* \leq Q_1$.*

Proof of Proposition 3.2 Part (a): If $r = -Q$, the inventory position under the (R,nQ) model is always less than or equal to zero so that the on-hand inventory is always zero, and hence, $V_{\mathrm{I}}(-Q,Q) = 0$. Then, following Proposition 3.1, we obtain $r_{\mathrm{v}}(Q) \geq -Q$.

Part (b): According to Federgruen and Zhang (1992), we know that

$$
-\sum_{y=r+1}^{r+Q} G(y)/Q - \mu C/Q
$$

is unimodal in r. Therefore,

$$
\pi(r,Q) = p - c - \sum_{y=r+1}^{r+Q} G(y)/Q - \mu C/Q
$$

is unimodal in r. As a result, by Proposition 3.1, we obtain Proposition 3.2(b).

[2] Notice that Proposition 3.1 is similar to Lemma 1 of Chen and Federgruen (2000).

Part (c): For any feasible solution (r, Q) of (P3.1) with $Q>Q_1$, by definition, we have $\pi(r,Q) \leq \pi_e(Q)<\pi_0$.

By Proposition 3.2, the following Algorithm 3.1 is derived to solve (P3.1):

Algorithm 3.1

00: (Initial step) set $r^* = -1$; $Q^* = 1$; and $\pi^* = (p-c-C)\mu - G(0)$;
01: set $\pi_0 = \pi^*$ and determine Q_1;
02: for $Q = 1$ to Q_1 do
03: begin
04:     set $r = -Q$;
05:     while $V_I(r+1,Q) \leq v$ and $r+1 \leq r_e(Q)$ do $r = r + 1$;
06:     if $\pi(r,Q) > \pi^*$ then
07:     begin
08:         set $r^* = r$; $Q^* = Q$; $\pi^* = \pi(r,Q)$;
09:         set $\pi_0 = \pi^*$ and update Q_1;
10:     end
11: end

The initial solution is $r = -1$ and $Q = 1$. The initial solution $(-1,\ 1)$ is feasible because $V_I(-1,1) = 0 < v$, and the corresponding $\pi(-1,\ 1) = (p-c-C)\mu - G(0)$. The loop specified in line 02 of Algorithm 3.1 follows Proposition 3.2(c); line 04 follows Proposition 3.2(a); line 05 follows Proposition 3.2(b); lines 06 to 10 update the solution if a better feasible solution, which induces a bigger $\pi(r,Q)$, is found.

3.3 Variance of Profit

The variance of profit measures directly the profit variation/uncertainty of an arbitrary period. So a retailer, which dislikes high variation/uncertainty in profit, can use the variance of profit as the "risk" measure under the (R, nQ) model.

Consider any period t in steady state. Let $V_p(r,Q)$ be the variance of the one-period profit. We consider the following mean-risk problem of the (R, nQ) model with the long-run average profit as the "mean" and the variance of one-period profit as the "risk" measure:

$$\max_{r,Q} \pi(r,Q)$$

$$\text{s.t. } V_p(r,Q) \leq w, \tag{P3.3}$$

where $w > 0$ is a prespecified maximum risk tolerance level (on the acceptable $V_p(r,Q)$). Besides the formulation given by (P3.3), we can also consider the following alternative mean-risk problem of the (R, nQ) model:

$$\begin{aligned}&\min_{r,Q} V_{\mathrm{p}}(r,Q)\\ &\text{s.t. } \pi(r,Q) \geq k,\end{aligned} \tag{P3.4}$$

where $k > 0$ is a prespecified minimum requirement of the long-run average of profit of the retailer. Consider specific $w = w_2>0$ and $k = k_2>0$, and suppose that (r_2, Q_2) is the optimal solution of (P3.3) for $w = w_2>0$, and $\pi(r_2, Q_2) \geq k_2>0$. For any $(r, Q) \neq (r_2, Q_2)$, $V_{\mathrm{p}}(r, Q)>V_{\mathrm{p}}(r_2, Q_2)$ if $\pi(r, Q)>\pi(r_2, Q_2) \geq k_2$. So (r_2, Q_2) is also the optimal solution of (P3.4) for $k = k_2>0$. Therefore, an optimal solution of (P3.3) for a given $w>0$ is also an optimal solution of (P3.4) for some $k > 0$, and vice versa. The pair of (w, k) is an efficient point in $\pi(r, Q) - V_{\mathrm{p}}(r, Q)$ space, i.e., an efficient point of the mean-risk problem of the (R, nQ) model with the long-run average profit as the "mean," and the variance of one-period profit as the "risk" measure. By varying the value of w, or k, in the above problems, we can obtain the efficient frontier. We note that w and k in the pair of (w, k) may not be a one-to-one mapping. In other words, a given $w > 0$ can pair up with different $k > 0$ to form an efficient point, and vice versa.

Let R be the ordering costs incurred at the beginning of period $t + l + 1$, and H be the holding and backorder costs incurred at the end of period $t + l$. Let $T = R + H$ be the total cost in a period. The one-period profit is $(p - c)D - T$. Since the expected value of the one-period profit is $\pi(r, Q)$, we have:

$$\begin{aligned}V_{\mathrm{p}}(r,Q) &= E[(p-c)D - R - H]^2 - [\pi(r,Q)]^2\\ &= (p-c)^2 E[D^2] + E[R^2] + E[H^2] - 2(p-c)E[DR]\\ &\quad -2(p-c)E[DH] - 2E[RH] - [\pi(r,Q)]^2\\ &= (p-c)^2(\sigma^2+\mu^2) + E[R^2] + E[H^2] - 2(p-c)E[DR]\\ &\quad -2(p-c)E[DH] - 2E[RH] - [\pi(r,Q)]^2.\end{aligned} \tag{3.5}$$

According to (3.5), to determine $V_{\mathrm{p}}(r, Q)$, we have to determine $E[R^2]$, $E[H^2]$, $E[DR]$, $E[DH]$, and $E[RH]$.

As $r < I(t + l) \leq r + Q$ and $I(t + l) = I(t) - D_l + mQ \equiv O[I(t) - D_l]$, where m represents the number of batches ordered in periods $t + 1, \ldots, t + l$, and m is the unique nonnegative integer so that $r < I(t + l) \leq r + Q$. The inventory position at the end of period $t + l$ is $I(t + l) - D$. If $I(t + l) - D > r$, then no order is placed at the beginning of period $t + l + 1$. If $r - nQ < I(t + l) - D \leq r - (n - 1)Q$, then nQ units are ordered, for any positive integer n. Since the ordering cost of each batch of product (Q units) is C, we have:

$$R = \sum_{n=0}^{\infty} (nC)\mathbf{1}(r - nQ < I(t+l) - D \leq r - (n-1)Q), \tag{3.6}$$

and

$$DR = \sum_{n=0}^{\infty} (nCD)\mathbf{1}(r - nQ < I(t+l) - D \leq r - (n-1)Q),$$

where $\mathbf{1}(\cdot)$ is the indicator function, i.e., $\mathbf{1}(x) = 1$ if statement x is true, and $\mathbf{1}(x) = 0$ if statement x is false. Since $I(t + l)$ is uniformly distributed on the interval $[r + 1, r + Q]$ in the steady state, and independent of D (see, e.g., Hadley and Whitin 1961), from (3.6), we have:

$$E[R] = \sum_{n=0}^{\infty} (nC) \Pr(r - nQ < I(t+l) - D \leq r - (n-1)Q)$$
$$= \sum_{n=0}^{\infty} \frac{nC}{Q} \sum_{y=r+1}^{r+Q} [F(y + nQ - r - 1) - F(y + (n-1)Q - r - 1)].$$

$$E[R^2] = \sum_{n=0}^{\infty} (nC)^2 \Pr(r - nQ < I(t+l) - D \leq r - (n-1)Q)$$
$$= \sum_{n=0}^{\infty} \frac{(nC)^2}{Q} \sum_{y=r+1}^{r+Q} [F(y + nQ - r - 1) - F(y + (n-1)Q - r - 1)]. \quad (3.7)$$

and

$$E[DR] = \sum_{n=0}^{\infty} \frac{nC}{Q} \sum_{y=r+1}^{r+Q} E[D\mathbf{1}(y + (n-1)Q - r \leq D < y + nQ - r)$$
$$= \sum_{n=0}^{\infty} \frac{nC}{Q} \sum_{y=r+1}^{r+Q} \left(\sum_{z=y-r+(n-1)Q}^{y+nQ-r-1} zf(z) \right). \quad (3.8)$$

Next, from the inventory balance equation (3.1), we have:

$$H = h(I(t) - D_{l+1})^+ + b(D_{l+1} - I(t))^+ \quad (3.9)$$

$$H = h(I(t) - D_l - D)^+ + b(D_l + D - I(t))^+. \quad (3.10)$$

By the fact that $(I(t) - D_{l+1})^+(D_{l+1} - I(t))^+ = 0$, and by using (3.9), we have:

$$H^2 = h^2[(I(t+l) - D_{l+1})^+]^2 + b^2[(D_{l+1} - I(t+l))^+]^2$$
$$+ bh(I(t+l) - D_{l+1})^+(D_{l+1} - I(t+l))^+$$
$$= h^2[(I(t+l) - D_{l+1})^+]^2 + b^2[(D_{l+1} - I(t+l))^+]^2.$$

and hence

$$E[H^2] = h^2 E[(I(t+l) - D_{l+1})^+]^2 + b^2 E[(D_{l+1} - I(t+l))^+]^2$$
$$= \frac{h^2}{Q} \sum_{y=r+1}^{r+1} \sum_{z=0}^{y} (y - z)^2 f_{l+1}(x) + \frac{b^2}{Q} \sum_{y=r+1}^{r+!} \sum_{z=y+1}^{\infty} (z - y)^2 f_{l+1}(x). \quad (3.11)$$

Next, from (3.6), (3.10), and the fact that $I(t), D_l$ and D are independent, we have:

$$E[RH] = \frac{C}{Q} \sum_{y=r+1}^{r+Q} \sum_{x=0}^{\infty} \sum_{z=0}^{\infty} n(y,x,z)[h(y-x-z)^+ + b(z+x-y)^+] f_L(x) f(z), \tag{3.12}$$

where $n(y,x,z)$ is the unique integer which satisfies

$$r < O[y-x] - z + n(y,z,x)Q \le r + Q.$$

Moreover, from (3.10) and the fact that $I(t), D_l$ and D are independent, we have:

$$E[DH] = \frac{1}{Q} \sum_{y=r+1}^{r+Q} \sum_{x=0}^{\infty} \sum_{z=0}^{\infty} [h(y-x-z)^+ + b(z+x-y)^+] f_L(x) f(z) \tag{3.13}$$

Having (3.7), (3.8), (3.11), (3.12) and (3.13), we can determine $V_p(r,Q)$.

Since $V_p(r,Q)$ is very complicated, we do not explore the analytical properties of it. Instead, we suggest a systematic searching method to solve (P3.3) or (P3.4) numerically.

3.4 Numerical Analysis

In this section, we consider examples that the demand D follows the binomial distribution with parameters 10 and 0.4, and the deliver lead time $l = 3$. Five sets of price-cost parameters are considered, which are given in Table 3.1. In particular, example 2 has a relatively big product cost c, example 3 has a relatively big backorder cost b, example 4 has a relatively big holding cost h, and example 5 has a relatively big ordering cost C. We explore the effects of different price-cost parameters on the efficient frontier.

Table 3.1 Parameters of the numerical examples

	Ordering cost, C	Retail price, p	Holding cost, h	Backorder cost, b	Production cost, c
Example 1	2	10	1	2	4
Example 2	2	10	1	2	7
Example 3	2	10	1	4	4
Example 4	2	10	3	2	4
Example 5	8	10	1	2	4

Table 3.2 Values of r and Q that maximize $\pi(r, Q)$

	r	Q	Maximum value of $\pi(r, Q)$
Example 1	2	5	20.1162
Example 2	2	5	8.1162
Example 3	3	5	19.4782
Example 4	1	4	18.3636
Example 5	0	10	17.14

We first determine the pair of (r, Q) that maximizes $\pi(r, Q)$ for each example, which is shown in Table 3.2. In particular, the maximum $\pi(r, Q)$ of example 2 is the smallest among the five examples because the unit product cost c is big. With a relatively big backorder cost b, example 3 has the biggest reorder point r in maximizing $\pi(r, Q)$. It is because a bigger r reduces the probability of having backorder. With a relatively big holding cost h, in contrast to example 3, example 4 has the smallest reorder point r in maximizing $\pi(r, Q)$. Moreover, example 4 has a smaller Q (compared to examples 1–3) as well. It is because both a smaller r and a smaller Q can decrease the on-hand inventory. With a relatively big ordering cost C, it is obvious that example 5 has the smallest r and the biggest Q among the five examples in maximizing $\pi(r, Q)$ because both a smaller r and a bigger Q decrease the number of reordering in a long run.

We next plot the efficient frontiers of the mean-risk problem of the (R, nQ) model. We first solve (P3.2) and (P3.4) for different values of k. After that, we trace out the efficient frontiers by plotting the non-inferior points on mean and risk, i.e., only the points that yield the minimum level of risk for a given level of expected profit, or the points that yield the maximum level of expected profit for a given risk level, are included in the efficient frontier. In example 1, $V_{\mathrm{I}}(-1, 1) = 0$ is the minimum variance of on-hand inventory. As $\pi(-1, 1) \geq 7$ and the maximum $\pi(r, Q) = 20.1162$, we only consider the range $7 \leq k < 20.1162$ in determining the optimal solutions of (P3.2). Similarly, as the variance of (one-period) profit of example 1 is minimized at $(r, Q) = (7, 9)$ and $\pi(7, 9) \geq 15$, we only consider the range $15 \leq k < 20.1162$ in determining the optimal solutions of (P3.4).

Figure 3.1 shows the result of plotting efficient frontiers of example 1. The optimal solutions of (P3.2) with different values of k are shown in Table 3.3, and the optimal solutions of (P3.4) with different values of k are shown in Table 3.5. From Table 3.3, we observe that $(r, Q) = (-1, 2)$ is the optimal solution for $k = 8$, 9, 10, 11, and 12 (and $(r, Q) = (-1, 3)$ is the optimal solution for $k = 13$, 14, and 15). Therefore, the same optimal solution can be obtained for different values of k. In plotting the efficient frontier, we focus on the optimal solutions and the associated mean and risk measures.

The efficient frontier of example 1 in $\pi(r, Q) - \sqrt{V_I(r, Q)}$ space, i.e., efficient frontiers of (P3.1) and (P3.2) are respectively shown in Table 3.4 and Fig. 3.1a, and the efficient frontiers of example 1 in $\pi(r, Q) - \sqrt{V_{\mathrm{P}}(r, Q)}$ space, i.e., efficient frontiers of (P3.3) and (P3.4) are respectively shown in Table 3.6 and Fig. 3.1b. As shown in Tables 3.5 and 3.6, on one hand, we find that the efficient points in the

Table 3.3 Optimal solutions of (P3.2) for different values of k

k	r	Q	$\pi(r,Q)$	$\sqrt{V_I(r,Q)}$
7	−1	1	8	0
8	*−1*	2	*12.9909*	*0.000026*
9	*−1*	2	*12.9909*	*0.000026*
10	*−1*	2	*12.9909*	*0.000026*
11	*−1*	2	*12.9909*	*0.000026*
12	*−1*	2	*12.9909*	*0.000026*
13	*−1*	*3*	*15.2749*	*0.000119*
14	*−1*	*3*	*15.2749*	*0.000119*
15	*−1*	*3*	*15.2749*	*0.000119*
16	−1	4	16.7914	0.000406
17	0	3	17.0552	0.000469
18	0	4	18.3399	0.001301
19	0	5	19.1309	0.003204
20	1	6	20.0706	0.015883
20.1162	2	5	20.1162	0.017398

Table 3.4 Efficient frontier of example 1 in $\pi(r,Q) - \sqrt{V_I(r,Q)}$ space

r	Q	$\pi(r,Q)$	$\sqrt{V_I(r,Q)}$
−1	1	8	0
−1	2	12.9909	0.000026
−1	3	15.2749	0.000119
−1	4	16.7914	0.000406
0	3	17.0552	0.000469
0	4	18.3399	0.001301
0	5	19.1309	0.003204
1	6	20.0706	0.015883
2	5	20.1162	0.017398

Table 3.5 Optimal solutions of (P3.4) for different values of k

k	r	Q	$\pi(r,Q)$	$\sqrt{V_P(r,Q)}$
15	7	9	15.1105	21.9041
16	6	9	16.1058	21.9043
17	5	9	17.0828	22.0138
18	4	9	18.0045	22.2764
19	2	10	19.0575	23.1483
20	2	6	20.0959	25.8352
20.1162	2	5	20.1162	27.1946

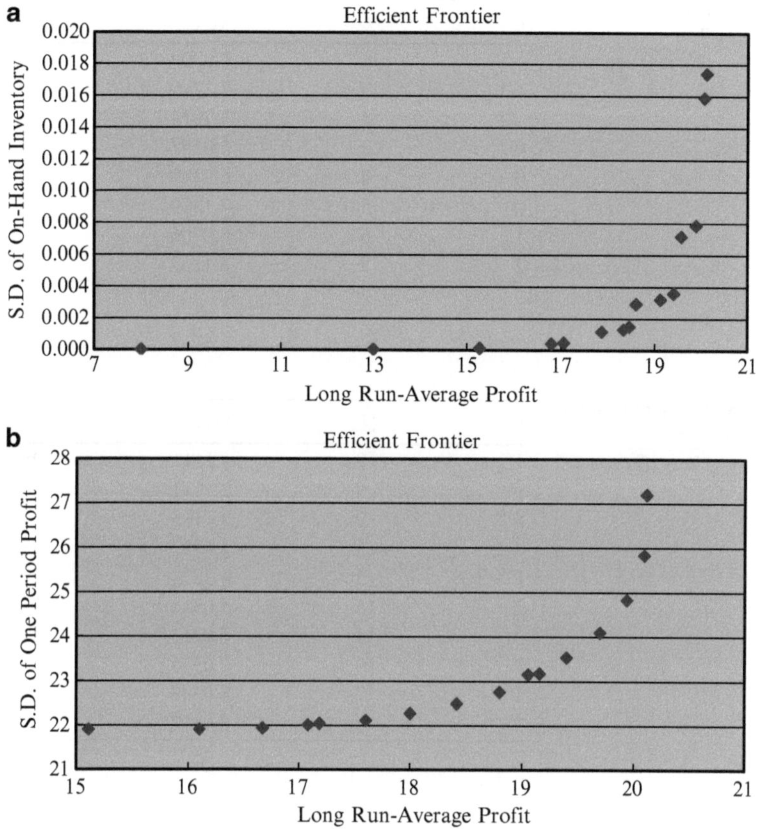

Fig. 3.1 (**a**) Efficient frontier of example 1 in $\pi(r,Q) - \sqrt{V_{\mathrm{I}}(r,Q)}$ space. (**b**) Efficient frontier of example 1 in $\pi(r,Q) - \sqrt{V_{\mathrm{P}}(r,Q)}$ space

Table 3.6 Efficient frontier of example 1 in $\pi(r,Q) - \sqrt{V_{\mathrm{P}}(r,Q)}$ space

r	Q	$\pi(r,Q)$	$\sqrt{V_{\mathrm{P}}(r,Q)}$
7	9	15.1105	21.9041
6	9	16.1058	21.9043
5	9	17.0828	22.0138
4	9	18.0045	22.2764
2	10	19.0575	23.1483
2	6	20.0959	25.8352
2	5	20.1162	27.1946

two spaces can be very different when $\pi(r,Q)$ is small. On the other hand, when $\pi(r,Q)$ is closer to its maximum, we observe that the differences among the efficient points of the two spaces become smaller. This observation can be explained by the fact that $\pi(r,Q)$ becomes dominating when it becomes big.

Table 3.7 Efficient frontier of example 2 in $\pi(r,Q) - \sqrt{V_I(r,Q)}$ space

r	Q	$\pi(r,Q)$	$\sqrt{V_I(r,Q)}$
−1	2	0.9909	0.000026
−1	3	3.2749	0.000119
−1	4	4.7914	0.000406
0	3	5.0552	0.000469
0	4	6.3399	0.001301
0	5	7.1309	0.003204
1	6	8.0706	0.015883
2	5	8.1162	0.017398

Table 3.8 Efficient frontier of example 2 in $\pi(r,Q) - \sqrt{V_P(r,Q)}$ space

r	Q	$\pi(r,Q)$	$\sqrt{V_P(r,Q)}$
10	9	1.1111	25.612891
9	9	2.1111	26.404056
8	9	3.1105	27.342079
4	15	3.4027	28.307874
3	15	4.2823	29.054535
2	15	5.0384	29.903006
3	11	6.0213	31.316224
2	10	7.0575	33.347616
1	7	8.0598	39.026333
2	5	8.1162	42.041683

Table 3.9 Efficient frontier of example 3 in $\pi(r,Q) - \sqrt{V_I(r,Q)}$ space

r	Q	$\pi(r,Q)$	$\sqrt{V_I(r,Q)}$
−1	2	5.9849	0.000026
−1	3	9.2359	0.000119
−1	4	11.6523	0.000406
0	3	12.8698	0.000469
0	4	14.8999	0.001301
0	5	16.2848	0.003204
1	5	18.2220	0.007839
3	5	19.4782	0.035540

The efficient frontiers of examples 2–5 are shown in Tables 3.7, 3.8, 3.9, 3.10, 3.11, 3.12, 3.13, 3.14 and Figs. 3.2, 3.3, 3.4, 3.5. From Tables 3.7 and 3.8, we find that the range of $\pi(r,Q)$ of the efficient frontier of example 2 is significantly smaller than other examples' because the unit production cost c in example 2 is bigger than the ones in other examples.

Table 3.10 Efficient frontier of example 3 in $\pi(r,Q)-\sqrt{V_{\mathrm{P}}(r,Q)}$ space

r	Q	$\pi(r,Q)$	$\sqrt{V_{\mathrm{P}}(r,Q)}$
18	1	1	23.381337
17	2	5.5	23.542045
17	3	6.333	23.621285
15	5	8.4	23.702136
13	6	10.1667	24.029772
10	9	12.	25.091659
8	9	14	27.191355
6	9	16	30.501087
4	8	18	36.319590
3	5	19.4782	45.309189

Table 3.11 Efficient frontier of example 4 in $\pi(r,Q)-\sqrt{V_{\mathrm{I}}(r,Q)}$ space

r	Q	$\pi(r,Q)$	$\sqrt{V_{\mathrm{I}}(r,Q)}$
−1	1	8	0
−1	2	12.9849	0.000026
−1	3	15.2359	0.000119
−1	5	17.5199	0.001164
1	4	18.3636	0.003582

Table 3.12 Efficient frontier of example 4 in $\pi(r,Q)-\sqrt{V_{\mathrm{P}}(r,Q)}$ space

R	Q	$\pi(r,Q)$	$\sqrt{V_{\mathrm{P}}(r,Q)}$
7	8	0.4988	26.881795
6	9	2.1022	27.136836
5	9	5.0640	27.412126
4	10	6.5401	27.508493
3	9	10.5990	27.554631
2	9	12.9214	27.627462
2	8	14.1615	27.705162
2	6	16.3821	28.335959
1	5	18.2220	30.096048
1	4	18.3636	31.781861

Table 3.13 Efficient frontier of example 5 in $\pi(r,Q)-\sqrt{V_{\mathrm{I}}(r,Q)}$ space

r	Q	$\pi(r,Q)$	$\sqrt{V_{\mathrm{I}}(r,Q)}$
−1	2	0.9909	0.000026
−1	3	7.2749	0.000119
−1	4	10.7914	0.000406
−1	5	13.0719	0.001164
−1	6	14.6091	0.002925
−1	8	16.3007	0.013755
0	10	17.14	0.085194

Table 3.14 Efficient frontier of example 5 in $\pi(r,Q) - \sqrt{V_P(r,Q)}$ space

r	Q	$\pi(r,Q)$	$\sqrt{V_P(r,Q)}$
15	3	0.3333	22.350040
15	4	2.5	22.487261
14	5	4.6	22.869613
13	6	6.1667	23.123024
11	9	8.4444	23.525915
9	11	10.0909	24.127179
5	17	12.1027	24.324730
3	16	14.3272	25.120101
1	15	16.0279	26.971235
0	10	17.14	32.781652

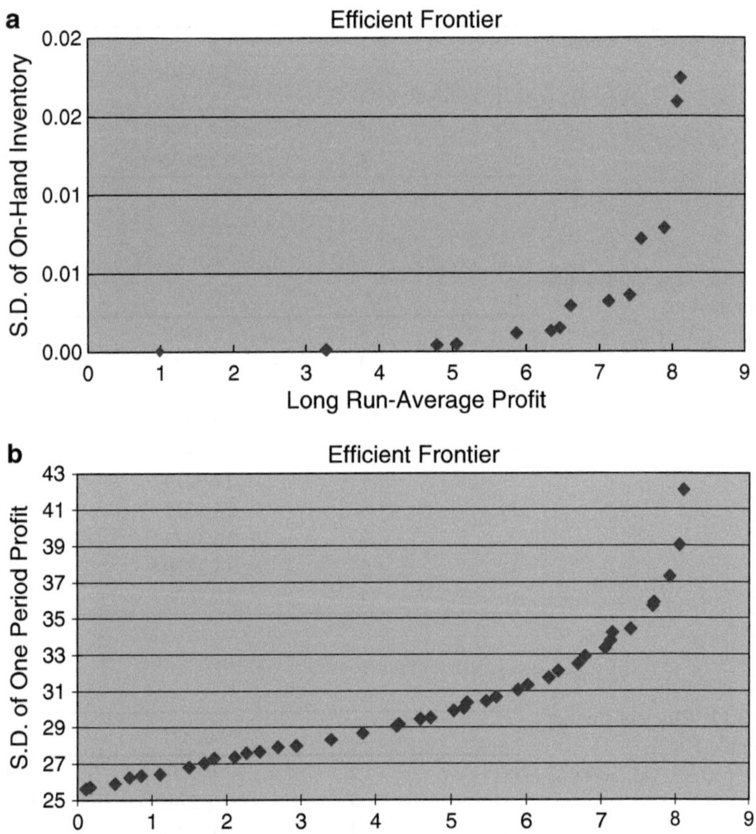

Fig. 3.2 (**a**) Efficient frontier of example 2 in $\pi(r,Q) - \sqrt{V_I(r,Q)}$ space. (**b**) Efficient frontier of example 2 in $\pi(r,Q) - \sqrt{V_P(r,Q)}$ space

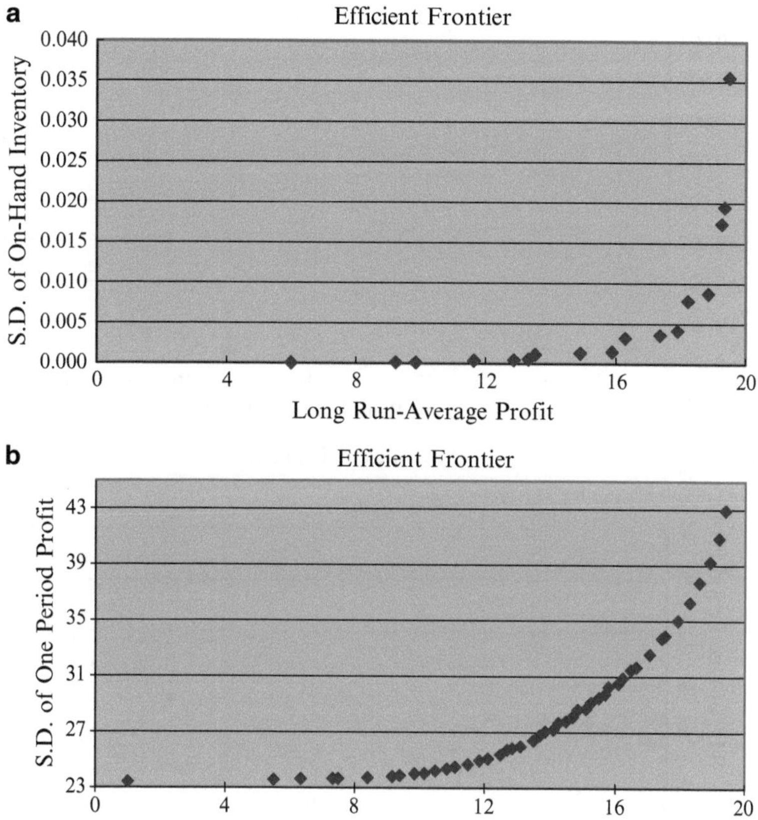

Fig. 3.3 (**a**) Efficient frontier of example 3 in $\pi(r, Q) - \sqrt{V_{\rm I}(r, Q)}$ space. (**b**) Efficient frontier of example 3 in $\pi(r, Q) - \sqrt{V_{\rm P}(r, Q)}$ space

We cannot find any obvious difference from Tables 3.9, 3.11, and 3.13. It means that the values of holding cost, backorder cost, and ordering cost have little effect on the efficient frontiers of (P3.1) and (P3.2), i.e., in $\pi(r, Q) - \sqrt{V_{\rm I}(r, Q)}$, for the numerical examples we considered.

Tables 3.10 and Table 3.12 show that there are big differences on the efficient points between example 3 and example 4 in the $\pi(r, Q) - \sqrt{V_{\rm P}(r, Q)}$ space. To be specific, the efficient reorder points r in example 3 are bigger than that in example 4. A bigger value of r reduces the number of backorder and increases the on-hand inventory. Therefore, in order to maintain the level of $\pi(r, Q)$: On one hand the retailer needs a bigger r to reduce the total backorder cost if the per unit backorder cost is big. On the other hand, the retailer needs a smaller r to reduce the total holding cost if the per unit holding cost is big.

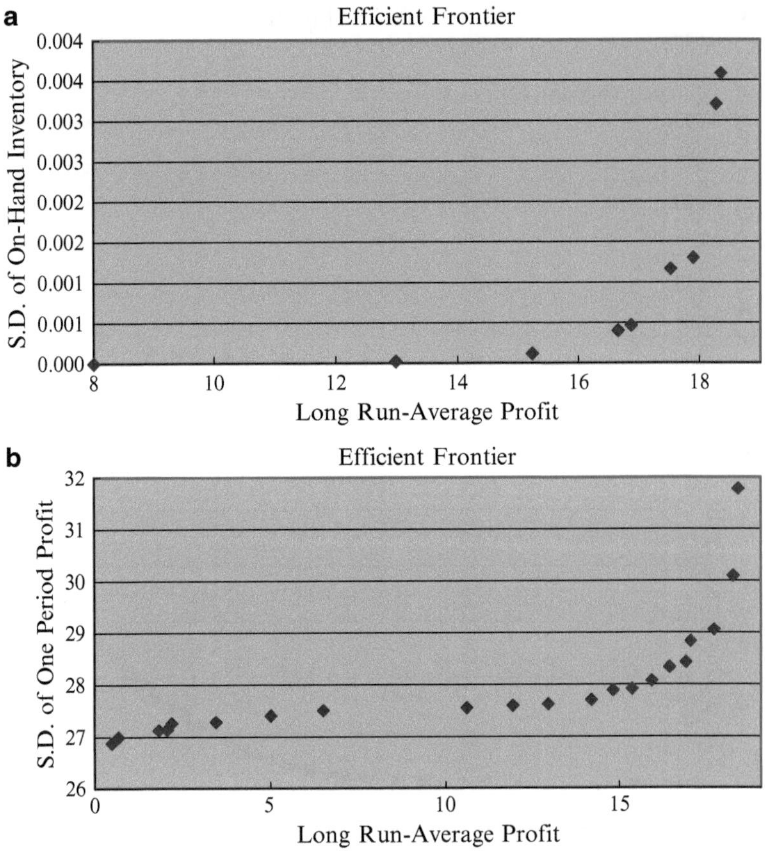

Fig. 3.4 (**a**) Efficient frontier of example 4 in $\pi(r,Q)-\sqrt{V_{\mathrm{I}}(r,Q)}$ space. (**b**) Efficient frontier of example 4 in $\pi(r,Q)-\sqrt{V_{\mathrm{P}}(r,Q)}$ space

From Tables 3.4, 3.8, 3.10, 3.12, and 3.14, we find that the efficient reorder points r in $\pi(r,Q)-\sqrt{V_{\mathrm{P}}(r,Q)}$ space are decreasing in $\pi(r,Q)$ as well as $\sqrt{V_{\mathrm{P}}(r,Q)}$. We observe these findings from the numerical examples, and it will be interesting to explore these findings analytically in future research.

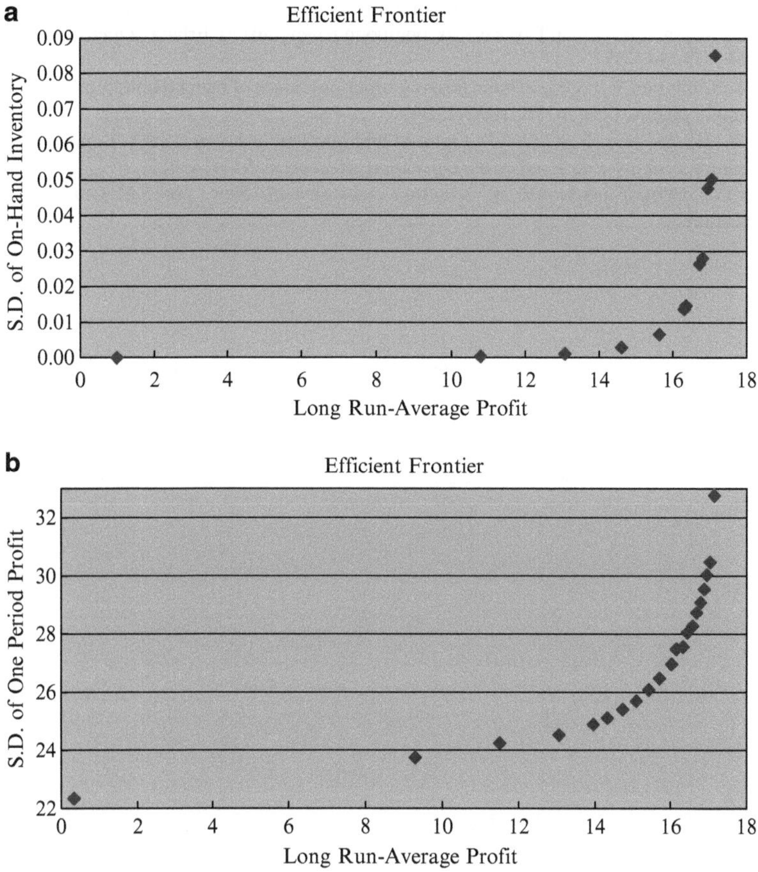

Fig. 3.5 (**a**) Efficient frontier of example 5 in $\pi(r,Q) - \sqrt{V_{\mathrm{I}}(r,Q)}$ space. (**b**) Efficient frontier of example 5 in $\pi(r,Q) - \sqrt{V_{\mathrm{P}}(r,Q)}$ space

References

Chen, F., & Federgruen, A. (2000). *Mean-variance analysis of basic inventory models*. Working paper, Columbia University.

Chen, F., & Zheng, Y. S. (1994). Evaluating echelon stock (R, nQ) policies in serial production/inventory systems with stochastic demand. *Management Science, 40*, 1262–1275.

Chen, F., & Zheng, Y. S. (1998). Near-optimal echelon-stock (R, nQ) policies in multistage serial systems. *Operations Research, 46*, 592–602.

Federgruen, A., & Zhang, Y. S. (1992). An efficient algorithm for computing an optimal (r, Q) policy in continuous review stochastic inventory systems. *Operations Research, 40*, 808–813.

Hadley, G., & Whitin, T. (1961). A family of inventory models. *Management Science, 7*, 351–371.

Lagodimos, A. G., Christou, I. T., & Skouri, K. (2012). Optimal (r, nQ, T) batch ordering with quantized supplies. *Computers and Operations Research, 39*, 259–268.

Larsen, C., & Kiesmüller, G. P. (2007). Developing a closed-form cost expression for an (R, s, nQ) policy where the demand process is compound generalized Erlang. *Operations Research Letters, 35*, 567–572.

Li, X., & Sridharan, V. (2008). Characterizing order processes of using (R, nQ) inventory policies in supply chains. *Omega, 36*, 1096–1104.

Shang, K. H., & Zhou, S. X. (2010). Optimal and heuristic echelon (r, nQ, T) policies in serial inventory systems with fixed costs. *Operations Research, 58*, 414–427.

Zipkin, P. (2000). *Foundations of inventory management*. New York: McGraw-Hill Higher Education.

Chapter 4
Mean-Risk Analysis of Supply Chain Coordination Problems

Supply chain coordination is an important topic in supply chain management. In the literature, under a stochastic demand, contracts such as returns contract (Pasternack 1985; Choi et al. 2008; Wu 2011), revenue sharing contract (Cachon and Lariviere 2005; Krishnan and Winter 2011; Sheu 2011), markdown money contract (Elmaghraby et al. 2008; Yin et al. 2009), quantity flexibility contract (Sethi et al 2004; Lian and Deshmukh 2009), and sales rebate contract (Taylor 2002; Arcelus et al. 2012) have been shown to be successful in coordinating a supply chain with risk neutral agents (see Cachon 2003; Tsay et al. 1999). Even though the studies with risk neutral agents can provide insights into supply chain coordination mechanisms, they lack precision in the sense that the agents could have different degrees of risk aversion. In this chapter, we consider a two-echelon supply chain with a single risk-neutral manufacturer and a single risk-averse retailer, and we study via a mean-risk analysis how target sales rebate (TSR) contract can help to coordinate a supply chain in this setting.[1] In the literature, there are several ways to define supply chain coordination with risk-averse supply chain agents (see Gan et al. 2004). Here, we adopt the definition of supply chain coordination as follows: *supply chain coordination is achieved when the expected profit of the supply chain is maximized.* We first apply the mean-risk approach to study the problem of supply chain coordination with a risk-averse retailer via TSR contracts. We then extend the mean-risk analysis (Chen and Federgruen 2000) to study the challenging case that includes sales effort-dependent demand.

---

[1] This chapter follows the work of Chiu et al. (2011a) on supply chain coordination with risk-averse agents. We acknowledge Elsevier to grant us the authorship right to revise and publish this work in a book format. In order to enhance the presentation of the mean-risk analysis on supply chain coordination problems, we consider a simplified model (cf. Chiu et al. 2011a) in this chapter.

T.-M. Choi and C.-H. Chiu, *Risk Analysis in Stochastic Supply Chains: A Mean-Risk Approach*, International Series in Operations Research & Management Science 178,
DOI 10.1007/978-1-4614-3869-4_4,

4.1 Supply Chain Model

We consider a two-echelon supply chain with a single risk-neutral manufacturer and a single risk-averse retailer. The retailer sells a newsboy type of fashionable product (Nahmias 1997) which is produced by the manufacturer. The retailer faces an uncertain market demand. The product demand follows a strictly positive density function $f(\cdot)$ and a cumulative distribution function $F(\cdot)$. We assume that there is a one-to-one mapping between $F(\cdot)$ and its argument. The unit retail price of the product is p. At the end of the selling season, any unsold product has zero salvage value. For the manufacturer, he bears a unit production cost of $c > 0$. Parameters p and c are exogenous parameters. We consider in this paper that the retailer's order can always be fulfilled by the manufacturer.

4.1.1 Risk-Averse Decision Model

We apply the mean-risk analysis to study the decision making problem of the risk-averse retailer. Specifically, we take the expected profit (EP) of the risk-averse retailer as the "mean" and the variance of profit (VP) of the risk-averse retailer as the "risk" measure. In other words, we adopt the *mean-variance* (MV) approach (Markowitz 1959) in this chapter to study the supply chain coordination problems with risk-averse agents.

Denote by $k > 0$ the expected profit target threshold of the risk-averse retailer. The decision making problem of the risk-averse retailer under the proposed MV model is defined as follows:

$$\begin{aligned} &\min \ V_{\mathrm{R}}(\cdot) \\ &\text{s.t. } E_{\mathrm{R}}(\cdot) \geq k \end{aligned} \tag{P4.1}$$

where $V_{\mathrm{R}}(\cdot)$ and $E_{\mathrm{R}}(\cdot)$ denote the variance of profit and the expected profit of the risk-averse retailer, respectively. Under problem (P4.1), the risk-averse retailer aims to minimize the variance of profit (the risk) under which the corresponding expected profit (the mean) is no less than the target minimum expected profit threshold k.

4.1.2 Supply Chain Coordination

We adopt the definition of supply chain coordination:

Definition 4.1 A contract θ coordinates a supply chain if there exists an autonomy decision y^{a} by members of the supply chain such that, for any decision y, $y^{\mathrm{a}} = \arg\ max_y(E_{\mathrm{SC}}(\theta, y))$ holds, where $E_{\mathrm{SC}}(\theta, y)$ is the supply chain's expected profit.

We adopt this definition because it is widely and commonly adopted in the literature, and maximizing $E_{SC}(\theta, y)$ gives an analytically tractable and meaningful benchmark for subsequent analysis. As a remark, there exist other interpretations or objectives on supply chain coordination, interested reader can refer to Gan et al. (2005) for some more discussions.

4.1.3 Supply Contracts

A wholesale price (WP) contract, denoted by $\theta_{WP}(w)$, is a supply contract that is offered by the manufacturer to the retailer in which the unit wholesale price w is the only contract parameter. To avoid trivial cases, we assume that $p > w > c > 0$.

A TSR is a payment from a manufacturer to a retailer based on the sales of the retailer to the end consumers (see Taylor 2002). A rebate $u > 0$ is paid for each unit sold beyond a sales target $t > 0$. A TSR contract, denoted by $\theta_{TSR}(w, u, t)$, is a supply contract that is offered by the manufacturer to the retailer in which w, u, and t are specified.

We consider that the manufacturer moves first and offers a supply contract to the retailer. After knowing the details of the contract, the retailer reacts and decides the order quantity.

A list of notations that are frequently used in this chapter is given in Table 4.1.

4.2 Structural Properties: EP and VP

We first study the structural properties of the EPs and VPs in the supply chain. For a given order quantity q, the profit, the EP, and the VP of the supply chain system are respectively given as follows:

$$\Pi_C(q) = (p-c)q + p\min(x-q, 0), \tag{4.1}$$

$$E_C(q) = (p-c)q - p\Psi(q), \text{ and} \tag{4.2}$$

$$V_C(q) = p^2\xi(0, q).$$

Notice that the specific details and type of supply contracts offered between the manufacturer and the retailer only affect the division of supply chain profit between them. They do not affect the supply chain's system profit, EP and VP.

Denote by q_C^* the order quantity that maximizes $E_C(q)$. It is easy to check that $E_C(q)$ is a concave function. Thus, q_C^* uniquely exists and it can be derived as follows by setting the first-order condition as zero.

$$q_C^* = F^{-1}[(p-c)/p].$$

Table 4.1 A list of notations

$\bar{F}(\cdot) = 1 - F(\cdot)$.
$\Psi(z) = \int_0^z F(x)\mathrm{d}x$, where $z \geq 0$.
$\Phi(y,z) = \int_y^z \bar{F}(x)\mathrm{d}x$, where $z > y \geq 0$.
$\Gamma(y,z) = \int_y^z xF(x)\mathrm{d}x$, where $z > y \geq 0$.
$\xi(y,z) = 2\{z\Psi(z) - \Psi(y)] - \Gamma(y,z)\} - [\Psi(z) - \Psi(y)]^2$, where $z > y \geq 0$.
$\omega(y,z) = (y-z)\Psi(y) + 2[z(\Psi(z) - \Psi(y)) - \Gamma(y,z)] - \Psi(z)[\Psi(z) - \Psi(y)]$, where $z > y \geq 0$.
$A'(\cdot)$: the first-order derivative of any arbitrary function $A(\cdot)$.
$A''(\cdot)$: the second-order derivative of any arbitrary function $A(\cdot)$.
Subscripts M, R, and C: the manufacturer, the retailer, and the supply chain, respectively.
Subscripts WP and TSR: the WP contract and the TSR contract, respectively.
EP: Expected profit.
VP: Variance of profit.
$\Pi_i(\cdot)$: profit of supply chain agent i, $i =$ M, R, and C.
$E_i(\cdot)$: expected profit of supply chain agent i, $i =$ M, R, and C.
$V_i(\cdot)$: variance of profit of supply chain agent i, $i =$ M, R, and C.

For a given $\theta_{\mathrm{WP}}(w)$, the profits of the retailer and the manufacturer are given by

$$\Pi_{\mathrm{R,WP}}(q) = (p-w)q + p\min(x-q,0) \text{ and}$$

$$\Pi_{\mathrm{S,WP}}(q) = (w-c)q,$$

respectively.

The expected profit and variance of profit of the manufacturer for a given $\theta_{\mathrm{WP}}(w)$ are respectively given by

$$E_{\mathrm{M,WP}}(q) = (w-c)q$$

and

$$V_{\mathrm{M,WP}}(q) = 0.$$

The expected profit and variance of profit of the retailer for a given $\theta_{\mathrm{WP}}(w)$ are

$$E_{\mathrm{R,WP}}(q) = (p-w)q - p\Psi(q) \text{ and} \tag{4.3}$$

$$V_{\mathrm{R,WP}}(q) = p^2\xi(0,q)$$

respectively.

For any given $\theta_{\mathrm{WP}}(w)$, the manufacturer's variance of profit is 0, and the variance of profit of the retailer equals the variance of the supply chain. In other words, the manufacturer bears no risk, but the retailer bears all the risk of the entire supply chain when the manufacturer offers a WP contract to the retailer.

As a simple projection from our analysis in Chap. 2, we have the following structural properties of EPs and VPs for a given $\theta_{\mathrm{WP}}(w)$: For any $q > 0$,

(1) $E_{\mathrm{R,WP}}(q)$and $E_{\mathrm{C}}(q)$ are concave functions of q
(2) $V_{\mathrm{R,WP}}(q)$ and $V_{\mathrm{C}}(q)$ are strictly increasing in q
(3) $E_{\mathrm{R,WP}}(q)$ is maximized at $q = q_{\mathrm{R,WP}}(w){:} = F^{-1}[(p-w)/p]$

Denote by $w_{\mathrm{WP}}(y)$ the wholesale price of a WP contract in which the maximum attainable retailer's EP equals $y \geq 0$. The value of $w_{\mathrm{WP}}(y)$ can be found by solving $h(w) = y$ for w, where

$$h(w) \equiv E_{\mathrm{R,WP}}(q_{\mathrm{R,WP}}(w)) = (p-w)q_{\mathrm{R,WP}}(w) - p\Psi(q_{\mathrm{R,WP}}(w)).$$

For any given $\theta_{\mathrm{TSR}}(w,u,t)$, the profits of the retailer and the manufacturer are

$$\Pi_{\mathrm{R,TSR}}(q) = (p-w)q + p\min(x-q,0) + u[q-t+\max(\min(x-q,0),t-q)], \tag{4.4}$$

and

$$\Pi_{\mathrm{M,TSR}}(q) = (w-c)q - u[q-t+\max(\min(x-q,0),t-q)],$$

respectively.

Since rebate is trivial if $q \leq t$, in the following discussion, we consider the case only when $q > t$. For $q > t$, the EPs of the retailer and the manufacturer for a given $\theta_{\mathrm{TSR}}(w,u,t)$ are given by

$$E_{\mathrm{R,TSR}}(q) = (p-w)q + p\Psi(q) + u\Phi(t,q) \text{ and} \tag{4.5}$$

$$E_{\mathrm{M,TSR}}(q) = (w-c)q - u\Phi(t,q),$$

respectively.

The VPs of the retailer and the manufacturer under TSR contract are given by

$$V_{\mathrm{R,TSR}}(q) = p^2\xi(0,q) + u^2\xi(t,q) + 2up\omega(t,q),$$

and

$$V_{\mathrm{M,TSR}}(q) = u^2\xi(t,q),$$

respectively.

Observe that, for any fixed $q > t, V_{R,TSR}(q) + V_{M,TSR}(q) > V_C(q)$. As $\Pi_{R,TSR}(q) + \Pi_{M,TSR}(q) = \Pi_C(q)$, the covariance of $\Pi_{R,TSR}(q)$ and $\Pi_{M,TSR}(q)$ is negative, i.e., they are negatively correlated. Next, from (4.3) and (4.5), the retailer's EP under TSR contract is

$$E_{R,TSR}(q) = \begin{cases} (p-w)q - p\Psi(q), & 0 < q \le t, \\ (p-w)q - p\Psi(q) + u\Phi(t,q), & q > t. \end{cases}$$

The first-order derivative of $E_{R,TSR}(q)$ with respect to q is

$$E_{R,TSR}{}'(q) = \begin{cases} p - w - pF(q), & 0 < q \le t, \\ p - w - pF(q) + u\bar{F}(q), & q > t. \end{cases}$$

We have:

$$\lim_{q \to t^-} E_{R,TSR}{}'(q) = p - w + pF(q),$$

and

$$\lim_{q \to t^+} E_{R,TSR}{}'(q) = p - w + pF(q) + u\bar{F}(q).$$

As $\lim_{q\to t^+} E_{R,TSR}{}'(q) > \lim_{q\to t^-} E_{R,TSR}{}'(q)$, $E_{R,TSR}(q)$ is non-differentiable at $q = t$ and $E_{R,TSR}(q)$ cannot be a concave function of q. The second-order derivative of $E_{R,TSR}(q)$ with respect to q is

$$E_{R,TSR}{}''(q) = \begin{cases} -pf(q) < 0, & 0 < q \le t, \\ -(p+u)f(q) < 0, & q > t. \end{cases}$$

Therefore $E_{R,TSR}(q)$ is piecewise concave in $0 < q < t$ and $q > t$.

Next, we consider the first-order derivatives of $V_{R,TSR}(q)$ and $V_{M,TSR}(q)$, with respect to q. We have:

$$V_{R,TSR}{}'(q) = 2(p+u)\bar{F}(q)\left[p\int_0^q F(x)\mathrm{d}x + u\int_t^q F(x)\mathrm{d}x\right] > 0,$$

and

$$V_{M,TSR}{}'(q) = 2u\bar{F}(q)\int_t^q F(x)\mathrm{d}x > 0.$$

Proposition 4.1 summarizes the findings on the structural properties of $\Pi_{R,TSR}(q)$, $\Pi_{M,TSR}(q)$, $E_{R,TSR}(q)$, $V_{R,TSR}(q)$, and $V_{M,TSR}(q)$.

Proposition 4.1

(a) $\Pi_{\mathrm{R,TSR}}(q)$ *and* $\Pi_{\mathrm{M,TSR}}(q)$ *are negatively correlated.*
(b) $E_{\mathrm{R,TSR}}(q)$ *is piecewise concave in* $0 < q < t$ *and* $q > t$.
(c) $V_{\mathrm{R,TSR}}(q)$ *and* $V_{\mathrm{M,TSR}}(q)$ *are strictly increasing in* q *for any* $q > t$.

Proposition 4.1(a) asserts that for a given $\theta_{\mathrm{TSR}}(w,u,t)$, if the retailer earns a bigger profit, then the manufacturer will earn a smaller profit, and vice versa. Hence, the manufacturer also bears risk for a given $\theta_{\mathrm{TSR}}(w,u,t)$. According to Proposition 4.1(b), there exists a unique local maximum point of $E_{\mathrm{R,TSR}}(q)$ for $0 < q < t$, and a unique local maximum point of $E_{\mathrm{R,TSR}}(q)$ for $q>t$. According to Proposition 4.1(c), a bigger order quantity q leads to bigger $V_{\mathrm{R,TSR}}(q)$ and $V_{\mathrm{S,TSR}}(q)$.

The order quantity that maximizes $E_{\mathrm{R,TSR}}(q)$ is given by

$$q^N_{\mathrm{R,TSR}} = \begin{cases} q_{\mathrm{R,TSR}}(w,u), & t<t_0,\\ q_{\mathrm{R,TSR}}(w,u) = q_{\mathrm{R,WP}}(w), & t=t_0,\\ q_{\mathrm{R,WP}}(w), & t>t_0, \end{cases} \tag{4.6}$$

where $q_{\mathrm{R,TSR}}(w,u) = F^{-1}[(p+u-w)/(p+u-s)]$, and t_0 is the unique solution of $g_0(t) = E_{\mathrm{R,TSR}}(q_{\mathrm{R,WP}}(w)) - E_{\mathrm{R,TSR}}(q_{\mathrm{R,TSR}}(w,u)) = 0$ (cf. Taylor 2002).

For a given $\theta_{\mathrm{WP}}(w)$, the optimal order quantity of a risk-averse retailer with $k \le E_{\mathrm{R,WP}}(q_{\mathrm{R,WP}}(w))$, which is denoted by $q^*_{\mathrm{R,WP}}$, satisfies

$$E_{\mathrm{R,WP}}(q^*_{\mathrm{R,WP}}) = k, \tag{4.7}$$

and

$$q^*_{\mathrm{R,WP}} \le q_{\mathrm{R,WP}}(w).$$

Notice that $q^*_{\mathrm{R,WP}}$ can be uniquely determined for any given $k \le E_{\mathrm{R,WP}}(q_{\mathrm{R,WP}}(w))$ (cf. Chap. 2).

The optimal order quantity of the risk-averse retailer for a given $\theta_{\mathrm{TSR}}(w,u,t)$ is more complicated. First of all, by (4.6) and the definition of $q_{\mathrm{R,WP}}(w)$,

(i) $E_{\mathrm{R,TSR}}(q) > E_{\mathrm{R,WP}}(q_{\mathrm{R,WP}}(w))$ is not attainable for any given $\theta_{\mathrm{TSR}}(w,u,t)$ with $t>t_0$.
(ii) $E_{\mathrm{R,TSR}}(q) > E_{\mathrm{R,TSR}}(q_{\mathrm{R,WP}}(w))$ is not attainable for any given $\theta_{\mathrm{TSR}}(w,u,t)$ with $t = t_0$.
(iii) $E_{\mathrm{R,TSR}}(q) > E_{\mathrm{R,TSR}}(q_{\mathrm{R,TSR}}(w,u))$ is not attainable for any given $\theta_{\mathrm{TSR}}(w,u,t)$ with $t > t_0$.

Therefore, we obtain Proposition 4.2:

Proposition 4.2

(a) If $t \ge t_0$, *then problem (P4.1) admits no solution for any* $k>E_{\mathrm{R,WP}}(q_{\mathrm{R,WP}}(w))$.
(b) If $t < t_0$, *then problem (P4.1) admits no solution for any* $k>E_{\mathrm{R,TSR}}(q_{\mathrm{R,TSR}}(w,u))$.

Actually, Proposition 4.2 shows the *feasible* range of k for which the optimal order quantity of a risk-averse retailer exists under different situations. Next, we derive the optimal order quantity of the risk-averse retailer for a given $\theta_{\mathrm{TSR}}(w,u,t)$ when the optimal solution of problem (P4.1) exists.

Proposition 4.3

(a) *For any given k that admits solution of the problem (P4.1), the optimal solution of problem (P4.1) is unique.*

(b) *The optimal solution of (P4.1), which is denoted by* $q^*_{\mathrm{R,TSR}}$, *satisfies*

(i) $E_{\mathrm{R,TSR}}(q^*_{\mathrm{R,TSR}}) = k,$ (4.8)

(ii) $q^*_{\mathrm{R,TSR}} \le q_{\mathrm{R,WP}}(w)$, if $t \ge t_0$, or $t_0 > t > q_{\mathrm{R,WP}}(w)$ and $k \le E_{\mathrm{R,WP}}(q_{\mathrm{R,WP}}(w))$,

(iii) $\hat{q} < q^*_{\mathrm{R,TSR}} \le q_{\mathrm{R,TSR}}(w,u)$, if $t_0 > t > q_{\mathrm{R,WP}}(w)$ and $k > E_{\mathrm{R,WP}}(q_{\mathrm{R,WP}}(w))$,

(iv) $q^*_{\mathrm{R,TSR}} \le q_{\mathrm{R,TSR}}(w,u)$, if $t < t_0$ and $t < q_{\mathrm{R,WP}}(w)$,

where $\hat{q}$ is the solution of $E_{\mathrm{R,TSR}}(q) = E_{\mathrm{R,TSR}}(q_{\mathrm{R,WP}}(w))$, for $q > q_{\mathrm{R,WP}}(w)$.

Proof of Proposition 4.3 For $t > t_0$, we know that $E_{\mathrm{R,WP}}(q)$ is concave in q, $E_{\mathrm{R,TSR}}(q) = E_{\mathrm{R,WP}}(q)$ for $q < t$, and $E_{\mathrm{R,TSR}}(q)$ is maximized at $q_{\mathrm{R,WP}}(w)$. Therefore, $E_{\mathrm{R,TSR}}(q)$ is strictly increasing in q, for $0 \le q \le q_{\mathrm{R,WP}}(w)$. Then, by the fact that $V_{\mathrm{R,TSR}}(q)$ is strictly increasing in q, $q^*_{\mathrm{R,TSR}}$ can be uniquely determined for any given k that admits solution of the problem (P4.1). Moreover, $q^*_{\mathrm{R,TSR}} \le q_{\mathrm{R,WP}}(w)$ and $q^*_{\mathrm{R,TSR}}$ satisfies $E_{\mathrm{R,TSR}}(q^*_{\mathrm{R,TSR}}) = k$.

Similar to the case for $t > t_0$, for $t = t_0$, $q^*_{\mathrm{R,TSR}} \le q_{\mathrm{R,WP}}(w)$ is unique and satisfies $E_{\mathrm{R,TSR}}(q^*_{\mathrm{R,TSR}}) = k$.

For $t_0 > t > q_{\mathrm{R,WP}}(w)$, $E_{\mathrm{R,TSR}}(q)$ is continuous in q, for $q \ge 0$, and is piecewise concave in $0 \le q \le t$ and in $q > t$, with the local maximum of $E_{\mathrm{R,TSR}}(q)$ at $q_{\mathrm{R,WP}}(w)$ and $q_{\mathrm{R,TSR}}(w,u)$. By (4.6), if $t_0 > t$, we have $E_{\mathrm{R,TSR}}(q_{\mathrm{R,WP}}(w)) < E_{\mathrm{R,TSR}}(q_{\mathrm{R,TSR}}(w,u))$. Then by the fact that $V_{\mathrm{R,TSR}}(q)$ is strictly increasing in q, $q^*_{\mathrm{R,TSR}}$ is unique, $\hat{q} < q^*_{\mathrm{R,TSR}} \le q_{\mathrm{R,TSR}}(w,u)$ and satisfies $E_{\mathrm{R,TSR}}(q^*_{\mathrm{R,TSR}}) = k$.

For $t \le q_{\mathrm{R,WP}}(w)$ and $t < t_0$, $E_{\mathrm{R,TSR}}(q)$ is strictly increasing in $0 \le q \le q_{\mathrm{R,TSR}}(w,u)$ ($E_{\mathrm{R,TSR}}{}'(q) > 0$ for $0 \le q < t$ and for $t < q \le q_{\mathrm{R,TSR}}(w,u)$, and $E_{\mathrm{R,TSR}}(q) > 0$ is continuous in q). Since $V_{\mathrm{R,TSR}}(q)$ is strictly increasing in q, $q^*_{\mathrm{R,TSR}} \le q_{\mathrm{R,TSR}}(w,u)$ is unique and satisfies $E_{\mathrm{R,TSR}}(q^*_{\mathrm{R,TSR}}) = k$.

Proposition 4.3 shows that the optimal order quantity of the risk-averse retailer heavily depends on the rebate target t. In particular, we note that

$$E_{\mathrm{R,WP}}(q_{\mathrm{R,WP}}(w)) = E_{\mathrm{R,TSR}}(q_{\mathrm{R,WP}}(w)),$$

for $t > t_0$, and

$$E_{\mathrm{R,WP}}(q_{\mathrm{R,WP}}(w)) = E_{\mathrm{R,TSR}}(q_{\mathrm{R,TSR}}(w,u)),$$

for $t = t_0$. Therefore, $q^*_{\mathrm{R,TSR}} = q^*_{\mathrm{R,WP}}$, which is given by (4.7), for $t \ge t_0$. This implies that the optimal solutions of (P4.1) are the same for given $\theta_{\mathrm{WP}}(w)$ and $\theta_{\mathrm{TSR}}(w,u,t)$ with the same w.

4.3 Supply Chain Coordination with Risk-Averse Agent

Assume that the manufacturer is the coordinator of the supply chain and aims to coordinate the supply chain by maximizing the supply chain's EP with the corresponding order quantity q_{C}^*. The retailer is risk-averse with an EP threshold k, and k is less than the maximum EP of the supply chain $E_{\mathrm{C}}(q_{\mathrm{C}}^*)$.

As q_{C}^* is a unique solution that maximizes $E_{\mathrm{C}}(q)$, to achieve supply chain coordination, the coordinator (manufacturer) needs to find a supply contract (here we use $\theta_{\mathrm{TSR}}(w,u,t)$) such that q_{C}^* is the unique optimal solution of (P4.1). In other words, the goal of coordination is to find a TSR contract $\theta_{\mathrm{TSR}}(w,u,t)$ such that $q_{\mathrm{R,TSR}}^* = q_{\mathrm{C}}^*$.

Proposition 4.4 *For any given $\theta_{\mathrm{TSR}}(w,u,t)$ with $c < w < p$, $u > 0$ and $t > 0$, $q_{R,TSR}^* = q_C^*$ if and only if w, u, and t satisfy the following conditions*

(i) $E_{\mathrm{R,TSR}}'(q_{\mathrm{C}}^*) \geq 0$ *and*
(ii) $E_{\mathrm{R,WP}}(q_{\mathrm{R,WP}}(w)) < k$.

Proof of Proposition 4.4 If $q_{\mathrm{R,TSR}}^* = q_{\mathrm{C}}^*$, then according to Proposition 4.2, $E_{\mathrm{R,TSR}}(q_{\mathrm{C}}^*) = k$. Next, we have $E_{\mathrm{R,WP}}(q) \leq E_{\mathrm{R,TSR}}(q)$ and $V_{\mathrm{R,TSR}}(q)$ is strictly increasing in q for any fixed q > 0. Since $q_{\mathrm{R,TSR}}^* > q_{\mathrm{R,WP}}(w)$ for p $>$ w $>$ c, we have $V_{\mathrm{R,TSR}}(q_{\mathrm{C}}^*) > V_{\mathrm{R,TSR}}(q_{\mathrm{R,WP}}(w))$ for p $>$ w $>$ c. Observe that q_{C}^* is the optimal order quantity of a risk-averse retailer, according to Proposition 4.2, we have $k = E_{\mathrm{R,TSR}}(q_{\mathrm{C}}^*) > E_{\mathrm{R,TSR}}(q_{\mathrm{R,WP}}(w))$. Therefore, $E_{\mathrm{R,WP}}(q_{\mathrm{R,WP}}(w)) < k$ and $E_{\mathrm{R,TSR}}'(q_{\mathrm{R,TSR}}^*) \geq 0$, or otherwise there exists $q' < q_{\mathrm{R,TSR}}^*$ such that $E_{\mathrm{R,TSR}}(q') > E_{\mathrm{R,TSR}}(q_{\mathrm{R,TSR}}^*)$ and $V_{\mathrm{R,TSR}}(q') > V_{\mathrm{R,TSR}}(q_{\mathrm{R,TSR}}^*)$. This contradicts to the fact that $q_{\mathrm{R,TSR}}^*$ is the optimal order quantity of a risk-averse retailer for a given $\theta_{\mathrm{TSR}}(w,u,t)$. Therefore, $E_{\mathrm{R,TSR}}'(q_{\mathrm{C}}^*) \geq 0$.

If $E_{\mathrm{R,TSR}}(q_{\mathrm{C}}^*) = k$ and $E_{\mathrm{R,WP}}(q_{\mathrm{R,WP}}(w)) < k$, then $E_{\mathrm{R,TSR}}(q_{\mathrm{C}}^*) = k > E_{\mathrm{R,WP}}(q_{\mathrm{R,WP}}(w))$. Note that, if $t \leq q_{\mathrm{R,WP}}(w)$, then $E_{\mathrm{R,TSR}}'(q) = p - w - (p-s)F(q) > p - w - (p-s)F(q_{\mathrm{R,WP}}(w)) \geq 0$, for $0 < q < t$. Since $E_{\mathrm{R,TSR}}(q)$ is concave in $q > t$ and it is maximized at $q_{\mathrm{R,TSR}}^*(w,u)$, so $E_{\mathrm{R,TSR}}(q)$ is strictly increasing in $q < q_{\mathrm{R,TSR}}(w,u)$. If $t > q_{\mathrm{R,WP}}(w)$, then $q_{\mathrm{R,WP}}(w)$ and $q_{\mathrm{R,TSR}}(w,u)$ are the local maxima of $E_{\mathrm{R,TSR}}(q)$. Next, since $E_{\mathrm{R,TSR}}(t) \leq E_{\mathrm{R,WP}}(q_{\mathrm{R,WP}}(w))$ for all $t > 0$, $E_{\mathrm{R,TSR}}(q_{\mathrm{C}}^*) \leq E_{\mathrm{R,WP}}(q_{\mathrm{R,WP}}(w))$ implies $q_{\mathrm{C}}^* > t$. Observe that $V_{\mathrm{R,TSR}}(q)$ is strictly increasing in q, and $E_{\mathrm{R,TSR}}(q)$ is either strictly increasing (for $t < q_{\mathrm{R,WP}}(w)$) or concave in $q > t$ (for $t < q_{\mathrm{R,WP}}(w)$). If $E_{\mathrm{R,TSR}}(q_{\mathrm{C}}^*) \geq 0$ and $E_{\mathrm{R,TSR}}(q_{\mathrm{C}}^*) > E_{\mathrm{R,WP}}(q_{\mathrm{R,WP}}(w))$, then we cannot find another q' such that $E_{\mathrm{R,TSR}}(q') > E_{\mathrm{R,TSR}}(q_{\mathrm{R,TSR}}^*)$ and $V_{\mathrm{R,TSR}}(q') > V_{\mathrm{R,TSR}}(q_{\mathrm{R,TSR}}^*)$. Therefore, $q_{\mathrm{C}}^* = q_{\mathrm{R,TSR}}^*$.

Proposition 4.4 shows the necessary and sufficient conditions for coordinating supply chain by $\theta_{\mathrm{TSR}}(w,u,t)$. As $E_{\mathrm{R,TSR}}'(q_{\mathrm{C}}^*) \geq 0$, we have $k \leq E_{\mathrm{R,TSR}}(q_{\mathrm{R,TSR}}(w,u))$. (In words, k is less than or equal to the maximum attainable retailer's EP for any given $\theta_{\mathrm{TSR}}(w,u,t)$ that achieves supply chain coordination). Moreover, $k > E_{\mathrm{R,WP}}(q_{\mathrm{R,WP}}(w))$ implies that the retailer's EP for a given $\theta_{\mathrm{TSR}}(w,u,t)$ that achieves supply chain coordination is bigger than the retailer's EP for a given $\theta_{\mathrm{WP}}(w)$ with the same w. Furthermore, Proposition 4.4 asserts that the expected profit of retailer under supply chain coordination is equal to k, and hence the expected profit of manufacturer under a coordinated supply chain is given by

$$E_{\mathrm{M,TSR}}(q_{\mathrm{C}}^*) = E_{\mathrm{C}}(q_{\mathrm{C}}^*) - k.$$

Next, we explore in more details the TSR contracts that can coordinate supply chain. Before presenting the results, we have the following terminologies. Let

$$\begin{aligned} g_1(t) &= t + \Psi(q_{\mathrm{C}}^*) - \Psi(t) - q_{\mathrm{C}}^* F(q_{\mathrm{C}}^*), \\ \hat{H}(t,\varphi,k) &= g_1(t)[k + p\Psi(q_{\mathrm{C}}^*)] - [E_{\mathrm{C}}(q_{\mathrm{C}}^*) - k + \varphi]\Phi(t, q_{\mathrm{C}}^*), \text{ and} \\ \tilde{H}(t,\varphi,k) &= \hat{H}(t,\varphi,k) - g_1(t)[p - w_{\mathrm{WP}}(k)]q_{\mathrm{C}}^*. \end{aligned}$$

Next, let
t_1 be the solution of $g_1(t) = 0$.
For any fixed $\varphi \geq 0$ and $k \geq 0$, let
$\hat{t}_{\varphi,k}$ be the solution of $\hat{H}(t,\varphi,k) = 0$ by solving t, and
$\tilde{t}_{\varphi,k}$ be the solution of $\tilde{H}(t,\varphi,k) = 0$ by solving t.

The properties of $\varphi,\ t_1,\ \hat{t}_{\varphi,k},\ \tilde{t}_{\varphi,k}$ are given in the following Proposition 4.5, and the physical meanings of $\varphi,\ t_1,\ \hat{t}_{\varphi,k},\ \tilde{t}_{\varphi,k}$ are discussed later in this chapter.

Proposition 4.5

(a) *A unique $t_1 \in (0, q_{\mathrm{C}}^*)$ exists, and for any given $0 < k < E_{\mathrm{C}}(q_{\mathrm{C}}^*)$, a unique $w_{\mathrm{WP}}(k)$ exists, and $c < w_{\mathrm{WP}}(k) < p$.*

(b) *For any given $\varphi \geq 0$ and $k < E_{\mathrm{C}}(q_{\mathrm{C}}^*)$, unique $\hat{t}_{\varphi,k}$ and $\tilde{t}_{\varphi,k}$ exist, and $t_1 < \hat{t}_{\varphi,k} < \tilde{t}_{\varphi,k} < q_{\mathrm{C}}^*$.*

(c) *For any given $\varphi \geq 0$, $k < E_{\mathrm{C}}(q_{\mathrm{C}}^*)$, and $z > y \geq 0$, we have $\hat{t}_{z,k} > \hat{t}_{y,k}$, and $\tilde{t}_{z,k} > \tilde{t}_{y,k}$.*

Proof of Proposition 4.5 Part (a). The first-order derivative of $g_1(t)$ with respect to t can be derived as follows

$$g_1{}'(t) = 1 - F(t) > 0,$$

for all $t > 0$. Thus, $g_1(t)$ is strictly increasing in t. Moreover,

$$g_1(0) = \Psi(q_{\mathrm{C}}^*) - q_{\mathrm{C}}^* F(q_{\mathrm{C}}^*) < 0,$$

and

$$g_1(q_{\mathrm{C}}^*) = q_{\mathrm{C}}^*(1 - F(q_{\mathrm{C}}^*)) > 0.$$

Therefore, a unique $t_1 \in (0, q_{\mathrm{C}}^*)$ exists. Next, for any given $k < E_{\mathrm{C}}(q_{\mathrm{C}}^*)$,

$$\begin{aligned} h'(w) &= [(p - w)q_{\mathrm{R,WP}}{}'(w) - pF(q_{\mathrm{R,WP}}(w))] - q_{\mathrm{R,WP}}(w) \\ &= -q_{\mathrm{R,WP}}(w) \\ &< 0. \end{aligned}$$

Thus, $h(w)$ is strictly decreasing in w. Moreover, $h(c) = E_{\mathrm{C}}(q_{\mathrm{C}}^*)$, and $h(p) = 0$. As a result, a unique $w_{\mathrm{WP}}(k)$ exists, and $c < w_{\mathrm{WP}}(k) < p$.

Part (b). The first-order derivative of $\hat{H}(t,\varphi,k)$ with respect to t is

$$\hat{H}'(t,\varphi,k) = (1-F(t))[p\Psi(q_C^*)+E_C(q_C^*)+\varphi] > 0.$$

Thus, $\hat{H}(t,\varphi,k)$ is strictly increasing in t. Moreover, we have:

$$\hat{H}(t_1,\varphi,k) = -[E_C(q_C^*)-k+\varphi]\Phi(t_1,q_C^*) < 0,$$

and

$$\hat{H}(q_C^*,\varphi,k) = q_C^*(1-F(q_C^*))[k+p\Psi(q_C^*)] > 0.$$

Therefore, a unique $\hat{t}_{\varphi,k}$ exists and $t_1 < \hat{t}_{\varphi,k} < q_C^*$. Similarly, $\tilde{H}'(t,\varphi,k) > 0$. Thus, $\tilde{H}(t,\varphi,k)$ is strictly increasing in t for any fixed $\varphi \geq 0$. Moreover, $\tilde{H}(q_C^*,\varphi,k) > 0$ and $\tilde{H}(\hat{t}_{\varphi,k},\varphi,k) < \hat{H}(\hat{t}_{\varphi,k},\varphi,k) = 0$. Therefore, a unique $\tilde{t}_{\varphi,k}$ exists and $t_1 < \hat{t}_{\varphi,k} < \tilde{t}_{\varphi,k} < q_C^*$.

Part (c). $\hat{t}_{z,k} > \hat{t}_{y,k}$ and $\tilde{t}_{z,k} > \tilde{t}_{y,k}$ because (1) $\hat{H}(t_1,\varphi,k)$ and $\tilde{H}(t,\varphi,k)$ are strictly decreasing in φ for $\varphi \geq 0$, i.e., $\hat{H}(t_1,z,k) < \hat{H}(t_1,y,k)$ and $\tilde{H}(t,z,k) < \tilde{H}(t,y,k)$ for any given t and any given z and y such that $z > y \geq 0$, and (2) $\hat{H}'(t,\varphi,k) > 0$ and $\tilde{H}'(t,\varphi,k) > 0$.

Proposition 4.6 *For any given* $u > 0$ *and* $0 < t < q_C^*$,

(a) $E_{R,TSR}(q_C^*) = k$ if and only if

$$w = p - [k + p\Psi(q_C^*) - u\Phi(q_C^*)]/q_C^*. \tag{4.9}$$

(b) $u\Phi(q_C^*) < k + p\Psi(q_C^*)$, and (4.9) holds if and only if $c < w < p$.
(c) $E_{R,TSR}'(q_C^*) \geq 0$ if and only if $ug_1(t) \geq E_C(q_C^*) - k$ holds.
(d) $ug_1(t) \geq E_C(q_C^*) - k$ implies $t > t_1$.
(e) $E_{R,WP}(q_{R,WP}(w)) < k$ if and only if (4.9) holds and

$$u\Phi(t,q_C^*) > k + p\Psi(q_C^*) - [p - w_{WP}(k)]q_C^*. \tag{4.10}$$

Proof of Proposition 4.6 By putting (4.9) into (4.5), we obtain part (a).

Part (b). By (4.9) and the fact that $c < w < p$, we have:

$$c < p - [k + p\Psi(q_C^*) - u\Phi(t,q_C^*)]/q_C^* < p.$$

Consider the first inequality, we have

$$u\Phi(t,q_C^*) > k + p\Psi(q_C^*) - (p-c)q_C^*, \tag{4.11}$$

and the second inequality is equivalent to

$$u\Phi(t,q_C^*) < k + p\Psi(q_C^*).$$

By (4.2), the left-hand side of (4.11) becomes

$$k - E_{\mathrm{C}}(q_{\mathrm{C}}^{*}) \leq 0.$$

If $u > 0$ and $t < q_{\mathrm{C}}^{*}$, then we have $\Phi(t, q_{\mathrm{C}}^{*}) > 0$. Therefore, inequality (4.11) must be satisfied.

Parts (c) and (d). $E_{\mathrm{R,TSR}}'(q_{\mathrm{C}}^{*}) = (p - w + u) - (p + u)F(q_{\mathrm{C}}^{*}) \geq 0$. By putting (4.9) into $E_{\mathrm{R,TSR}}'(q_{\mathrm{C}}^{*})$, and then rearranging the terms, we obtain

$$ug_1(t) \geq E_{\mathrm{C}}(q_{\mathrm{C}}^{*}) - k.$$

Since $u > 0$ and $E_{\mathrm{C}}(q_{\mathrm{C}}^{*}) - k > 0$, we have $ug_1(t) \geq E_{\mathrm{C}}(q_{\mathrm{C}}^{*}) - k$ implies $g_1(t) > 0$, i.e.,

$$ug_1(t) < 0 < E_{\mathrm{C}}(q_{\mathrm{C}}^{*}) - k,$$

which leads to a contradiction. As $g_1(t)$ is strictly increasing in t and $g_1(t_1) = 0$, $g_1(t) > 0$ if and only if $t > t_1$.

Part (e). On one hand, if (4.9) and (4.10) hold, then we have $w > w_{\mathrm{WP}}(k)$, and hence

$$E_{\mathrm{R,WP}}(q_{\mathrm{R,WP}}(w)) = h(w) < h(w_{\mathrm{WP}}(k)) = k.$$

On the other hand, $k > E_{\mathrm{R,WP}}(q_{\mathrm{R,WP}}(w))$ for $w > w_{\mathrm{WP}}(k)$. We thus obtain (4.10) by (4.9).

Proposition 4.6 provides all the conditions for $\theta_{\mathrm{TSR}}(w, u, t)$ to coordinate the supply chain. We note that the "coordinating" w depends on u and t. The inequalities in (b), (c) and (e) of Proposition 4.6 include u and t but exclude w. Therefore, to determine $\theta_{\mathrm{TSR}}(w, u, t)$ that coordinates supply chain, we need to find u and t that satisfy Propositions 4.6(b) and (c) at first and then Proposition 4.6(e). After that, w can be determined according to (4.9).

Proposition 4.7 *For any given $k < E_{\mathrm{C}}(q_{\mathrm{C}}^{*})$, $\theta_{\mathrm{TSR}}(w, u, t)$ coordinates supply chain if and only if there exists $\varphi \geq 0$ such that*

(i) $\hat{t}_{\varphi,k} < t < \tilde{t}_{\varphi,k}$
(ii) $u = [E_{\mathrm{C}}(q_{\mathrm{C}}^{*}) - k + \varphi]/g_1(t)$
(iii) w is given by (4.9)

Proof of Proposition 4.7 Condition (iii) directly follows from (4.9). When $\varphi \geq 0$, condition (ii) is equivalent to

$$ug_1(t) \geq E_{\mathrm{C}}(q_{\mathrm{C}}^{*}) - k.$$

When u satisfies condition (ii) and $\varphi \geq 0$,

$$u\Phi(t, q_{\mathrm{C}}^{*}) < k + p\Psi(q_{\mathrm{C}}^{*})$$

is equivalent to $t > \hat{t}^*_{\varphi,k}$, and (4.10) is equivalent to $t < \tilde{t}^*_{\varphi,k}$. By this result, together with $t > \hat{t}^*_{\varphi,k} > t_1$, we obtain $ug_1(t) \geq E_C(q^*_C) - k,\ t > t_1$, and (4.10) holds. By Proposition 4.6, the $\theta_{TSR}(w, u, t)$ that satisfies all the conditions of Proposition 4.7 achieves supply chain coordination, and it is the only form of the TSR contract that can coordinate the supply chain.

According to Proposition 4.7, there are multiple t that satisfy condition (i) of Proposition 4.7, for some $\varphi \geq 0$. Observe that conditions (i) and (ii) of Proposition 4.7 require a specific φ. Therefore, in determining $\theta_{TSR}(w, u, t)$ for coordinating supply chain, the manufacturer has to decide φ first. By having φ, the manufacturer selects t by following condition (i) of Proposition 4.7, and then u can be determined according to condition (ii) of Proposition 4.7. Finally, w can be calculated according to (4.9). Therefore, φ is an internal decision variable of the manufacturer. Denote by $\theta^{I}_{TSR}(w, u, t, \varphi)$, an internally specified TSR contract of the manufacturer. We note that $\theta^{I}_{TSR}(w, u, t, \varphi)$ is only for the internal use of the manufacturer, and the retailer knows $\theta_{TSR}(w, u, t)$ only. Furthermore, Proposition 4.7 also shows that if any two of w, u, and t are fixed, there exist multiple $\theta_{TSR}(w, u, t)$ that can coordinate supply chain because there are an infinite number of φ such that $\varphi \geq 0$.

Proposition 4.8 *For given* $k < E_C(q^*_C)$ *and* $0 < t < q^*_C$, *if* $\theta^{I}_{TSR}(w, u, t, \varphi)$ *coordinates supply chain, then the associated retailer's variance of profit is strictly increasing in* ϕ.

Proof of Proposition 4.8 For any given $k < E_C(q^*_C)$, if $\theta^{I}_{TSR}(w, u, t, \varphi)$ coordinates supply chain, then $t < q^*_C$ and

$$\mathrm{d}V_{R,TSR}(q)/\mathrm{d}\varphi = 2u\xi(t, q^*_C) + 2p\omega(t, q^*_C)/g_1(t).$$

For any $y < z$, we have

$$\xi(y,z) \geq \int_y^z (z + y - 2x)F(x)\mathrm{d}x > F(y)\int_y^z (z + y - 2x)F(x)\mathrm{d}x = 0,$$

and

$$\omega(y,z) \geq 2\int_y^z (z - x)F(x)\mathrm{d}x - \left(\int_y^z F(x)\mathrm{d}x\right)^2 \geq 0.$$

Therefore, $\mathrm{d}V_{R,TSR}(q)/\mathrm{d}\varphi > 0$.

According to Proposition 4.8, the value of φ directly affects the level of the retailer's variance of profit. Therefore, φ is the risk level indicator of the retailer. Specifically, when the manufacturer offers $\theta^{I}_{TSR}(w, u, t, \varphi)$ to the retailer to coordinate supply chain, the retailer bears a higher risk level with a bigger φ. As φ is an internal decision variable of the manufacturer, the manufacturer can control the risk level of the retailer by adjusting φ.

4.4 Numerical Analyses

To supplement the analytical findings derived above, we conduct numerical analyses in this section. We consider an example with the following parameters: Demand follows a uniform distribution with a lower bound $a = 0$, and a upper bound $b = 1,000$. The unit retail selling price $p = 50$, the unit product cost $c = 20$. With these parameters, the order quantity that maximizes the expected profit of supply chain is given by $q_C^* = 600$, and the associated expected profit of supply chain is given by $E_C(q_C^*) = 9,000$. As the expected profit target threshold of the risk-averse retailer k must be smaller than the maximum attainable supply chain's expected profit $E_C(q_C^*)$, or otherwise, supply chain coordination can never be achieved. We thus only consider the case with $k <$ 9,000 in the numerical analysis.

Following Proposition 4.7, we have the following steps to derive the TSR contract $\theta_{TSR}(w, u, t)$ that coordinates supply chain:

Step (1). Decide the value of φ.

Step (2). Calculate $\hat{t}_{\varphi,k}$ and $\tilde{t}_{\varphi,k}$, namely the lower bound and the upper bound of t for achieving supply chain coordination for given k and φ.

Step (3). Select the sales target t such that

$$\hat{t}_{\varphi,k} < t < \tilde{t}_{\varphi,k}.$$

Step (4). Calculate the value of the rebate u such that

$$u = [E_C(q_C^*) - k + \varphi]/g_1(t).$$

Step (5). Calculate the value of the wholesale price w such that

$$w = p - [k + p\Psi(q_C^*) - u\Phi(q_C^*)]/q_C^*.$$

We consider eight values of k = 1,000, 2,000, 3,000, 4,000, 5,000, 6,000, 7,000, 8,000. Table 4.2 shows TSR contracts with $\varphi = 0$ that can coordinate the supply chain for all k = 1,000, 2,000, ..., 8,000. As $\hat{t}_{0,k} < 400 < \tilde{t}_{0,k}$, for all k = 1,000, 2,000, ..., 8,000, we select $t = 400$ for all k = 1,000, 2,000, ..., 8,000. We can observe that when $t = 400$ is fixed, the associated u and w for supply chain coordination are decreasing in k.

According to Proposition 4.7, $\varphi \geq 0$ and $t \in (\hat{t}_{\varphi,k}, \tilde{t}_{\varphi,k})$ can be selected freely in determining the TSR for coordinating the supply chain. Next, we fix k = 5,000 and consider different values of φ and t, separately, to explore the effects of having different values of φ and t on supply chain coordination.

We consider six values of $\varphi = 0, 100, 200, 300, 400, 500$, to study the effects of having different values of φ on supply chain coordination. Table 4.3 shows TSR contracts with $t = 400$ that can coordinate the supply chain. As $\hat{t}_{\varphi,5,000} < 400 <$

Table 4.2 TSR contracts with $\varphi = 0$ that coordinate supply chain for different values of k

k	$\hat{t}_{\varphi,k}$	$\tilde{t}_{\varphi,k}$	t	u	w
1,000	346.8	434.3	400	57.14	42.86
2,000	326.7	464.4	400	50	40
3,000	307.2	488.7	400	42.86	37.14
4,000	288.2	510.1	400	35.71	34.29
5,000	269.7	529.8	400	28.57	31.43
6,000	251.7	548.3	400	21.43	28.57
7,000	234.1	566.0	400	14.29	25.71
8,000	216.8	583.2	400	7.14	22.86

Table 4.3 TSR contracts with $t = 400$ and different values of φ that coordinate supply chain with $k = 5{,}000$

φ	$\hat{t}_{\varphi,k}$	$\tilde{t}_{\varphi,k}$	t	u	w	$V_C(q_C^*)$
0	269.7	529.8	400	28.57	31.43	12,397
100	271.1	531.2	400	29.29	31.55	12,460
200	272.5	532.5	400	30	31.67	12,522
300	273.9	533.9	400	30.71	31.79	12,584
400	275.3	535.0	400	31.43	31.90	12,647
500	276.6	536.2	400	32.14	32.02	12,709

Table 4.4 $\theta_{\text{TSR}}(w, u, t)$ for supply chain coordination with $k = 5{,}000$, $\varphi = 0$ and different values of t

t	u	w	$V_C(q_C^*)$
280	65.79	46.32	18,996
320	45.05	38.02	15,401
360	34.72	33.89	13,548
400	28.57	31.43	12,397
440	24.51	29.80	11,603
480	21.65	28.66	11,019
520	19.53	27.81	10,572

$\tilde{t}_{\varphi,5{,}000}$, for all $\varphi = 0, 100, 200, \ldots, 500$, we select $t = 400$ for all $\varphi = 0, 100, 200, \ldots, 500$. When $k = 500$ and $t = 400$ are fixed, we find that (1) $0 < \hat{t}_{\varphi,k} < \hat{t}_{\varphi,k} < q_C^*$, and $\hat{t}_{\varphi,k}$ and $\tilde{t}_{\varphi,k}$ are increasing with φ, which is consistent with Proposition 4.5; (2) u and w are increasing with φ; and (3) the associated variance of profit of the retailer is increasing with φ, which is consistent with Proposition 4.8, i.e., for t is fixed, the retailer bears a higher risk level with a bigger φ.

As $\hat{t}_{\varphi,k} = 269.7$ and $\tilde{t}_{\varphi,k} = 529.77$ for $k = 5{,}000$ and $\varphi = 0$, we consider seven values of $t = 280$, 320, 360, 400, 440, 480, and 520 to study the effects of having different values of t on supply chain coordination. Table 4.4 shows the TSR contracts with $\varphi = 0$ and different values of t that can coordinate the supply chain with $k = 5{,}000$. When we fix $k = 5{,}000$ and $\varphi = 0$, we find that u, w, and the associated variance of profit of the retailer are decreasing with φ. Therefore, the retailer bears a smaller risk level with a bigger t, for $k = 5{,}000$ and $\varphi = 0$.

4.5 Coordination with Sales Effort-Dependent Demand

In this section, we consider the scenario when product demand depends on the amount of sales effort,

$$D(e) = ex,$$

where $e \geq 0$ is the level of sales effort in selling the product, and x is a random variable with a distribution function $F(x)$. The cost for the retailer to exert e units of effort is $C(e)$. Similar to (Taylor 2002), we assume that $C(e)$ is convex, strictly increasing and $C(0) = 0$. Therefore, the marginal cost of the sales effort is increasing. We assume that quantity and effort decisions are both decided prior to observing the state of market demand. For more discussions on the use of this kind of multiplicative effort-demand model, refer to Taylor (2002).

Define $Q = q/e$. The profit and the expected profit of the supply chain in the function of Q and e are given below:

$$\Pi_C(Q, e) = e[(p - c)Q + p\min(x - Q, 0)] - C(e),$$

and

$$E_C(Q, e) = e[(p - c)Q + p\Psi(Q)] - C(e),$$

respectively. Denote by (Q_C^*, e_C^*) and $q_C^* = Q_C^* e_C^*$, respectively, the optimal joint decisions and the optimal order quantity of the supply chain, which maximize $E_C(Q, e)$. Moreover, let $\pi_C^* = E_C(Q_C^*, e_C^*)$.

Proposition 4.9

(a) *The optimal joint decisions* (Q_C^*, e_C^*) *that maximize* $E_C(Q, e)$ *are unique and satisfy*

(i) $F(Q_C^*) = (p - c)/p$ *and*
(ii) $C'(e_C^*) = (p - c)Q_C^* - p\Psi(Q_C^*)$. (4.12)

(b) $\pi_C^* = e_C^*[(p - c)Q_C^* - p\Psi(Q_C^*)] - C(e_C^*)$.

Proof of Proposition 4.9 The first-order and second-order partial derivatives of $E_C(Q, e)$ with respect to Q

are

$$\partial E_C(Q, e)/\partial Q = e[p - c + pF(Q)],$$

and

$$\partial^2 E_C(Q, e)/\partial Q^2 = -epf(Q) < 0,$$

respectively. Therefore, for any fixed $e \geq 0$, $E_C(Q, e)$ is concave in Q and the optimal Q that maximizes $E_C(Q, e)$ is $Q_C^* = F^{-1}((p - c)/p)$. We note that Q_C^* is independent of e for any $e \geq 0$. By putting Q_C^* into $E_C(Q,e)$ and then considering the first-order and second-order derivatives of $E_C(Q_C^*, e)$ with respect to e, we have

$$E_C'(Q_C^*, e) = (p-c)Q_C^* - p\Psi(Q_C^*) - C'(e),$$

and

$$E_C''(Q_C^*, e) = -C''(e).$$

Therefore, $\Pi_C(Q_C^*, e)$ is concave in e, and e_C^* satisfies (4.12).

As (Q_C^*, e_C^*) is the unique solution that maximizes $E_C(Q,e)$, to achieve supply chain coordination, the coordinator (manufacturer) needs to find a supply contract (here we use TSR contract $\theta_{\text{TSR}}(w,u,t)$) such that (Q_C^*, e_C^*) is the unique optimal solution of (P4.1), i.e., the (Q_C^*, e_C^*) is the optimal joint decisions of the retailer for a given $\theta_{\text{TSR}}(w,u,t)$.

Notice that TSR contract is the same as the wholesale-price-only contract if $t \geq q$, we hence consider only the case of $q > t$ in the analysis. Moreover, the retailer's expected profit cannot exceed the supply chain's expected profit. Thus, we confine our analysis to the case with

$$k < \pi_C^*.$$

Similarly, for any given $\theta_{\text{WP}}(w)$, the retailer's profit and expected profit functions, respectively, are derived as follows:

$$\Pi_{\text{R,WP}}(Q,e) = e[(p-w)Q + p\min(x-Q,0)] - C(e),$$

and

$$E_{\text{R,WP}}(Q,e) = e[(p-w)Q - p\Psi(Q)] - C(e).$$

The variance of profit of retailer for given Q, e, and $\theta_{\text{WP}}(w)$ is

$$V_{\text{R,WP}}(Q,e) = e^2p^2\xi(0,Q).$$

Denote by $(Q_{\text{R,WP}}(w), e_{\text{R,WP}}(w))$ the joint decisions of the retailer that maximize $E_{\text{R,WP}}(Q,e)$, and let $q_{\text{R,WP}}(w) = Q_{\text{R,WP}}(w)e_{\text{R,WP}}(w)$. Moreover, let $\pi_{\text{R,WP}}(w) = E_{\text{R,WP}}(Q_{\text{R,WP}}(w), e_{\text{R,WP}}(w))$ be the maximum attainable expected profit of the retailer for a given $\theta_{\text{WP}}(w)$.

Similar to the optimal joint decisions of the supply chain, $(Q_{\text{R,WP}}(w), e_{\text{R,WP}}(w))$ must satisfy

(i) $F(Q_{\text{R,WP}}(w)) = (p-w)/p$

(ii) $C'(e_{\text{R,WP}}(w)) = (p-w)Q_{\text{R,WP}}(w) - p\Psi(Q_{\text{R,WP}}(w))$ (4.13)

(iii) $\pi_{\text{R,WP}}(w) = e_{\text{R,WP}}(w)C'(e_{\text{R,WP}}(w)) - C(e_{\text{R,WP}}(w))$

As $w > c$, we have $q_{\mathrm{R,WP}}(w) < q^*_{\mathrm{C}}$, $e_{\mathrm{R,WP}}(w) < e^*_{\mathrm{C}}$ and $\pi_{\mathrm{R,WP}}(w) < \pi^*_{\mathrm{C}}$. For any given $\theta_{\mathrm{WP}}(w)$, the maximum attainable retailer's profit is $\pi_{\mathrm{R,WP}}(w)$. Hence, for any given $\theta_{\mathrm{WP}}(w)$, problem (P4.1) admits no feasible solution if $k > \pi_{\mathrm{R,WP}}(w)$.

Proposition 4.10

(a) $e_{\mathrm{R,WP}}(w)$ *and* $\pi_{\mathrm{R,WP}}(w)$ *are decreasing in* w *for* $c < w < p$.

(b) $V_{\mathrm{R,WP}}(Q,e)$ *is strictly increasing in* $Q \geq 0$ *for any fixed* $e > 0$, *and is strictly increasing in* $e \geq 0$ *for any fixed* $Q > 0$.

Proof of Proposition 4.10 Part (a). We have:

$$Q_{\mathrm{R,WP}}{}'(w) = -[pf(Q_{\mathrm{R,WP}}(w))]^{-1} < 0.$$

By taking the first-order derivative on both the sides of (4.13) with respect to w, we obtain

$$\begin{aligned} e_{\mathrm{R,WP}}{}'(w) &= \frac{-1}{C''(e_{\mathrm{R,WP}}(w))}\left[Q_{\mathrm{R,WP}}(w) + \frac{(p-w) - pF(Q_{\mathrm{R,WP}}(w))}{pf(Q_{\mathrm{R,WP}}(w))}\right] \\ &= \frac{-Q_{\mathrm{R,WP}}(w)}{C''(e_{\mathrm{R,WP}}(w))} < 0. \end{aligned}$$

By considering the first-order derivative of $\pi_{\mathrm{R,WP}}{}'(w)$ with respect to w, we obtain

$$\pi_{\mathrm{R,WP}}{}''(w) = e_{\mathrm{R,WP}}(w)C''(e_{\mathrm{R,WP}}(w))e_{\mathrm{R,WP}}{}'(w) < 0.$$

Therefore, $e_{\mathrm{R,WP}}(w)$ and $\pi_{\mathrm{R,WP}}(w)$ are strictly decreasing in w.

Part (b). We have:

$$\partial V_{\mathrm{R,WP}}(Q,e)/\partial e = 2ep^2\xi(0,Q) > 0$$

for all $Q > 0$, and

$$\partial V_{\mathrm{R,WP}}(Q,e)/\partial Q = e^2p^2(1-F(Q))\Psi(Q) > 0$$

for all $e > 0$.

Proposition 4.10(a) implies that, for any given $\theta_{\mathrm{WP}}(w)$, retailer's sales effort and retailer's maximum attainable expected profit are decreasing in w. Moreover, as $\pi_{\mathrm{R,\,WP}}(c) = \pi^*_{\mathrm{C}}$ and $\pi_{\mathrm{R,\,WP}}(p) = 0$, there exists a unique $\tilde{w} \in (c,p)$ such that $\pi_{\mathrm{R,WP}}(\tilde{w}) = k$.

According to Proposition 4.10(b),

$$V_{\mathrm{R,WP}}(Q,e) > V_{\mathrm{R,WP}}(Q_{\mathrm{R,WP}}(w), e_{\mathrm{R,WP}}(w)),$$

for any $Q > Q_{\mathrm{R,WP}}$ and $e > e_{\mathrm{R,WP}}(w)$. Therefore, $q^*_{\mathrm{R,WP}} < q_{\mathrm{R,WP}}(w)$ and $e^*_{\mathrm{R,WP}} < e_{\mathrm{R,WP}}(w)$. Hence, the wholesale pricing only contract fails to coordinate supply chain when both quantity and sales effort are decisions.

Define (i) $T = t/e$ and $T_C^* = t/e_C^*$, (ii) $\alpha(w,u) = u(1-F(Q_C^*)) - w + c$, and (iii) $\beta(w,u,t) = u[\Phi(T_C^*,Q_C^*) + T_C^*(1-F(T_C^*))] - (w-c)q_C^*$. For any given $\theta_{TSR}(u,w,t)$ and $q > t$, the retailer's profit is

$$\Pi_{R,TSR}(Q,e) = e\{(p-w)Q + p\min(x-Q,0) + u[Q - T + \max(\min(x-Q,0), T-Q)]\}.$$

The expectation and the variance of $\Pi_{R,TSR}(Q,e)$ can be respectively derived as follows

$$E_{R,TSR}(Q,e) = e[(p-w)Q + p\Psi(Q) + u\Phi(T,Q)] - C(e),$$

and

$$V_{R,TSR}(Q,e) = e^2[p^2\xi(x,Q) + u^2\xi(T,Q) + 2up\omega(x-Q,0)].$$

Proposition 4.11
(a) *For any fixed* $e \geq 0$, $E_{R,TSR}(Q,e)$ *is concave in* Q *for* $Q > 0$, *and* $V_{R,TSR}(Q,e)$ *is strictly increasing in* Q *for* $Q > 0$.
(b) *For any fixed* $Q \geq 0$, $V_{R,TSR}(Q,e)$ *is strictly increasing in* e *for* $e > 0$.

Proof of Proposition 4.11 The second-order partial derivative of $E_{R,TSR}(Q,e)$ with respect to Q is

$$\partial^2 E_{R,TSR}(Q,e)/\partial Q^2 = -(p+u)f(Q) < 0.$$

Therefore, $E_{R,TSR}(Q,e)$ is concave in Q for any fixed $e \geq 0$. The first-order partial derivative of $V_{R,TSR}(Q,e)$ with respect to Q is

$$\begin{aligned}\partial V_{R,TSR}(Q,e)/\partial Q = {} & 2e[p^2(1-F(Q))\Psi(Q) \\ & + u^2(1-F(Q))[\Psi(Q) - \Psi(T)] \\ & + up[F(Q)\Phi(T,Q) + (1-F(Q))\Psi(Q) \\ & + 2(\Psi(Q) - \Psi(T))] \\ & > 0.\end{aligned}$$

The first-order partial derivative of $V_{R,TSR}(Q,e)$ with respect to e is

$$\begin{aligned}\partial V_{R,TSR}(Q,e)/\partial e = {} & 2\{e^{-1}V_{R,TSR}(Q,e) + uT[uF(T)\Phi(T,Q) \\ & + p[2(Q-T)F(T) + (1-F(T))\Psi(T)]\} \\ & > 0.\end{aligned}$$

Proposition 4.11 shows the structural properties of $E_{R,TSR}(Q,e)$ and $V_{R,TSR}(Q,e)$. Next, we go back to the supply chain coordination problem with risk-averse retailer and sales effort.

Proposition 4.12 *If* $\theta_{\mathrm{TSR}}(w,u,t)$ *coordinates supply chain, then the following hold*

(i) $w > \tilde{w}$
(ii) $E_{\mathrm{R,TSR}}(Q_{\mathrm{C}}^*, e_{\mathrm{C}}^*) = k$
(iii) $\alpha(w,u) \geq 0$ and $\beta(w,u,t) > 0$

Proof of Proposition 4.12 For a given $\theta_{\mathrm{TSR}}(w,u,t)$, and given that $Q = Q_{\mathrm{C}}^*$ and $e = e_{\mathrm{C}}^*$, we have:

$$e_{\mathrm{C}}^*\alpha(w,u) = \partial E_{\mathrm{R,TSR}}(Q,e)/\partial Q,$$

$$\beta(w,u,t) = \partial E_{\mathrm{R,TSR}}(Q,e)/\partial w,$$

$$2(e_{\mathrm{C}}^*)^2 B_Q(w,u,t) = \partial V_{\mathrm{R,TSR}}(Q,e)/\partial Q,$$

$$2B_e(w,u,t) = \partial V_{\mathrm{R,TSR}}(Q,e)/\partial e,$$

where

$$\begin{aligned} B_e(w,u,t) = {} & uTp[2(Q_{\mathrm{C}}^* - T_{\mathrm{C}}^*)F(T_{\mathrm{C}}^*) + (1 - F(T_{\mathrm{C}}^*)\Psi(T_{\mathrm{C}}^*)] \\ & + u^2 TF(T_{\mathrm{C}}^*)\Phi(T_{\mathrm{C}}^*, Q_{\mathrm{C}}^*) + \frac{V_{\mathrm{T,TSR}}(Q_{\mathrm{C}}^*, e_{\mathrm{C}}^*)}{e_{\mathrm{C}}^*}, \end{aligned}$$

and

$$\begin{aligned} B_Q(w,u,t) = {} & (1 - F(Q_{\mathrm{C}}^*))[p^2\Psi(Q_{\mathrm{C}}^*) + u^2(\Psi(Q_{\mathrm{C}}^*) - \Psi(T_{\mathrm{C}}^*))] \\ & + up[(1 - F(Q_{\mathrm{C}}^*))(2\Psi(Q_{\mathrm{C}}^*) - \Psi(T_{\mathrm{C}}^*))] + (Q_{\mathrm{C}}^* - T_{\mathrm{C}}^*)F(Q_{\mathrm{C}}^*) \\ & + \Psi(Q_{\mathrm{C}}^*) - \Psi(T_{\mathrm{C}}^*). \end{aligned}$$

According to Proposition 4.10 and by the definition of $\tilde{w}$, we know that $w < \tilde{w}$ and $\pi_{\mathrm{R,WP}}(w) > k$ are equivalent. As $V_{\mathrm{R,WP}}(Q,e) < V_{\mathrm{R,TSR}}(Q,e)$ for all $Q > 0$ and $e > 0$, and $V_{\mathrm{R,WP}}(Q,e) < V_{\mathrm{R,TSR}}(Q_{\mathrm{C}}^*, e_{\mathrm{C}}^*)$ for all $Q < Q_{\mathrm{C}}^*$ and $e < e_{\mathrm{C}}^*$ (according to Proposition 4.10), the optimal joint decisions of retailer are not $(Q_{\mathrm{C}}^*, e_{\mathrm{C}}^*)$ for any given $\theta_{\mathrm{TSR}}(w,u,t)$, and hence $\theta_{\mathrm{TSR}}(w,u,t)$ fails to coordinate the supply chain. As a result, condition (i) of Proposition 4.12 is the necessary condition of supply chain coordination.

According to (P4.1), if $\theta_{\mathrm{TSR}}(w,u,t)$ coordinates the supply chain with risk-averse retailer, then $E_{\mathrm{R,TSR}}(Q_{\mathrm{C}}^*, e_{\mathrm{C}}^*) \geq k$.

Notice that $E_{\mathrm{R,TSR}}(Q,e)$ is continuous at $Q = Q_{\mathrm{C}}^*$ and $e = e_{\mathrm{C}}^*$. According to Proposition 4.11 (i.e., $V_{\mathrm{R,WP}}(Q,e)$ is strictly increasing in Q, for any fixed $e \geq 0$, and $V_{\mathrm{R,WP}}(Q,e)$ is strictly increasing in e, for any fixed $Q \geq 0$), there exist some $Q < Q_{\mathrm{C}}^*$ and $e < Q_{\mathrm{C}}^*$ such that $E_{\mathrm{R,TSR}}(Q_{\mathrm{C}}^*, e_{\mathrm{C}}^*) \geq k$ and $V_{\mathrm{R,TSR}}(Q,e) < V_{\mathrm{R,TSR}}(Q_{\mathrm{C}}^*, e_{\mathrm{C}}^*)$

if $E_{\mathrm{R,TSR}}(Q_{\mathrm{C}}^{*}, e_{\mathrm{C}}^{*}) > k$. In other words, $Q = Q_{\mathrm{C}}^{*}$ and $e = e_{\mathrm{C}}^{*}$ are not the optimal joint decisions for the retailer. Therefore, $E_{\mathrm{R,TSR}}(Q_{\mathrm{C}}^{*}, e_{\mathrm{C}}^{*}) = k$ if $\theta_{\mathrm{TSR}}(w, u, t)$ coordinates the supply chain (condition (ii) of Proposition 4.12).

According to Proposition 4.11(a), $V_{\mathrm{R,TSR}}(Q, e)$ is strictly increasing in Q, for any fixed $e \geq 0$. If $\alpha(w, u) < 0$, then there exist some $Q < Q_{\mathrm{C}}^{*}$ such that

$$E_{\mathrm{R,TSR}}(Q_{\mathrm{C}}^{*}, e_{\mathrm{C}}^{*}) < E_{\mathrm{R,TSR}}(Q, e_{\mathrm{C}}^{*})$$

and

$$V_{\mathrm{R,TSR}}(Q_{\mathrm{C}}^{*}, e_{\mathrm{C}}^{*}) < V_{\mathrm{R,TSR}}(Q, e_{\mathrm{C}}^{*})$$

Hence, $Q = Q_{\mathrm{C}}^{*}$ and $e = e_{\mathrm{C}}^{*}$ are not optimal to the retailer, and $\theta_{\mathrm{TSR}}(w, u, t)$ fails to coordinate the supply chain. Similarly, according to Proposition 4.11(b), $V_{\mathrm{R,TSR}}(Q, e)$ is strictly increasing in e, for any fixed $Q \geq 0$. If $\beta(w, u, t) < 0$, then there exist some $e < e_{\mathrm{C}}^{*}$ such that

$$E_{\mathrm{R,TSR}}(Q_{\mathrm{C}}^{*}, e_{\mathrm{C}}^{*}) < E_{\mathrm{R,TSR}}(Q_{\mathrm{C}}^{*}, e)$$

and

$$V_{\mathrm{R,TSR}}(Q_{\mathrm{C}}^{*}, e_{\mathrm{C}}^{*}) < V_{\mathrm{R,TSR}}(Q_{\mathrm{C}}^{*}, e)$$

Hence, $Q = Q_{\mathrm{C}}^{*}$ and $e = e_{\mathrm{C}}^{*}$ are not optimal to the retailer, and $\theta_{\mathrm{TSR}}(w, u, t)$ fails to coordinate supply chain. Last, if $\beta(w, u, t) = 0$, then

$$\alpha(w, u)Q_{\mathrm{C}}^{*} = u[Q_{\mathrm{C}}^{*}(1 - F(Q_{\mathrm{C}}^{*})) - T_{\mathrm{C}}^{*}(1 - F(T_{\mathrm{C}}^{*})) - \Phi(T_{\mathrm{C}}^{*}, Q_{\mathrm{C}}^{*})] < 0, \quad \forall Q > T,$$

which violates the condition of $\alpha(w, u) \geq 0$.

Proposition 4.12 shows the necessary conditions under which $\theta_{\mathrm{TSR}}(w, u, t)$ achieves supply chain coordination. As $w > \tilde{w}$ implies that $\pi_{\mathrm{R,WP}}(\tilde{w}) > \pi_{\mathrm{R,WP}}(w)$, and by definition $\pi_{\mathrm{R,WP}}(\tilde{w}) = k$, conditions (i) and (ii) of Proposition 4.12 imply that $E_{\mathrm{R,TSR}}(Q_{\mathrm{C}}^{*}, e_{\mathrm{C}}^{*}) > \pi_{\mathrm{R,WP}}(w)$.

Moreover, according to condition (ii) of Proposition 4.12, the expected profit of the retailer under supply chain coordination is given by $E_{\mathrm{R,TSR}}(Q_{\mathrm{C}}^{*}, e_{\mathrm{C}}^{*}) = k$. As the expected profit of the supply chain under supply chain coordination is given by π_{C}^{*}, the expected profit of the manufacturer under supply chain coordination is given by $\pi_{\mathrm{C}}^{*} - k$.

Furthermore, given that $(Q, e) = (Q_{\mathrm{C}}^{*}, e_{\mathrm{C}}^{*})$, we have

$$\partial E_{\mathrm{R,TSR}}(Q, e)/\partial Q = e_{\mathrm{C}}^{*}\alpha(w, u),$$

and

$$\partial E_{\mathrm{R,TSR}}(Q,e)/\partial e = \beta(w,u,t).$$

Therefore, according to condition (iii) of Proposition 4.12, $E_{\mathrm{R,TSR}}(Q_{\mathrm{C}}^*, e_{\mathrm{C}}^*) = k$ is less than the maximum retailer's attainable expected profit for a given $\theta_{\mathrm{TSR}}(w,u,t)$ that coordinates the supply chain.

Proposition 4.13 *Suppose that $\theta_{\mathrm{TSR}}(w,u,t)$ can coordinate the supply chain.*

(a) If $\alpha(w,u) > 0$, *then*

$$e_{\mathrm{C}}^*\alpha(w,u)B_e(w,u,t) = \beta(w,u,t)B_Q(w,u,t).$$

(b) If $\alpha(w,u) = 0$, *then*

$$e_{\mathrm{C}}^*\alpha(w,u)B_e(w,u,t) \geq \beta(w,u,t)B_Q(w,u,t).$$

Proof of Proposition 4.13 For any fixed $k \in (0, \pi_{\mathrm{C}}^*)$, $E_{\mathrm{R,TSR}}(Q,e) = k$ implies that

$$u[Q(1-F(Q_{\mathrm{C}}^*)) - w + c)]\mathrm{d}Q = \{u[\Phi(T,Q) + uT(1-F(T))] - (w-c)Q\}de. \tag{4.14}$$

Moreover,

$$\mathrm{d}V_{\mathrm{R,TSR}}(Q,e) = [\partial V_{\mathrm{R,TSR}}(Q,e)/\partial e]\mathrm{d}Q - [\partial V_{\mathrm{R,TSR}}(Q,e)/\partial Q]\mathrm{d}e.$$

If $\alpha(w,u) > 0$, then there exist (Q_i, e_i), where $i = 1{,}2$, and $Q_1 < Q_{\mathrm{C}}^* < Q_2$ and $e_1 > e_{\mathrm{C}}^* > e_2$, such that

$$E_{\mathrm{R,TSR}}(Q_i, e_i) = k, \quad \forall i = 1,2.$$

Therefore, if $\theta_{\mathrm{TSR}}(w,u,t)$ can coordinate the supply chain, it implies that $\mathrm{d}V_{\mathrm{R,TSR}}(Q,e)/de = 0$ for $(Q,e) = (Q_{\mathrm{C}}^*, e_{\mathrm{C}}^*)$. By putting (4.14) into $\mathrm{d}V_{\mathrm{R,TSR}}(Q,e)$ and considering $\mathrm{d}V_{\mathrm{R,TSR}}(Q,e)/de = 0$ for $(Q,e) = (Q_{\mathrm{C}}^*, e_{\mathrm{C}}^*)$, we obtain Proposition 4.13(a).

Part (b). As $E_{\mathrm{R,TSR}}(Q,e)$ is concave in Q for any fixed $e \geq 0$, if $\alpha(w,u) = 0$, then there exist (Q_i, e_i), where $i = 1$, 2, and $Q_1 < Q_{\mathrm{C}}^* < Q_2$ and $e_1, e_2 > e_{\mathrm{C}}^*$, such that

$$E_{\mathrm{R,TSR}}(Q_i, e_i) = k, \quad \forall i = 1,2.$$

Therefore, if $\theta_{\mathrm{TSR}}(w,u,t)$ is the supply chain coordinating contract, it means that $\mathrm{d}V_{\mathrm{R,TSR}}(Q,e) \geq 0$ for $(Q,e) = (Q_{\mathrm{C}}^*, e_{\mathrm{C}}^*)$. By putting (4.14) into $\mathrm{d}V_{\mathrm{R,TSR}}(Q,e)$and considering $\mathrm{d}V_{\mathrm{R,TSR}}(Q,e) \geq 0$ for $(Q,e) = (Q_{\mathrm{C}}^*, e_{\mathrm{C}}^*)$, we obtain Proposition 4.13(b).

Proposition 4.13 supplements Proposition 4.12. To be specific, the two conditions of Proposition 4.13 ensure that $V_{\mathrm{R,TSR}}(Q_{\mathrm{C}}^*, e_{\mathrm{C}}^*) < V_{\mathrm{R,TSR}}(Q, e)$ for all $(Q,e) \neq (Q_{\mathrm{C}}^*, e_{\mathrm{C}}^*)$ that satisfies $E_{\mathrm{R,TSR}}(Q,e) = k$ for different situations. For $\alpha(w,u) > 0$, the retailer can keep $E_{R,TSR}(Q,e)=k$ by varying e and Q simultaneously from (Q_C^*, e_C^*), while for $\alpha(w,u)=0$, the retailer must increase e to keep $E_{R,TSR}(Q,e)=k$ when the retailer deviates Q from Q_C^*.

Based on Proposition 4.12, we can focus on the case for $w>\tilde{w}$ and $E_{R,TSR}(Q,e)=k$ to determine the TSR contract that can coordinate supply chain with risk-averse retailer and sales effort.

Proposition 4.14 *For any given $\theta_{TSR}(w,u,t)$ with $p>w>\tilde{w}$, $u>0$, and $q_C^*>t>0$, $\theta_{TSR}(w,u,t)$ will coordinate the supply chain if and only if $(Q,e)=(Q_C^*,e_C^*)$ is the unique solution of the following problem*

$$\begin{aligned}&\min_{(Q,e)\in\Theta(w,u,t)} V_{R,TSR}(Q,e)\\&\text{s.t. } E_{R,TSR}(Q,e)=k.\end{aligned} \tag{P4.2}$$

Proof of Proposition 4.14 If $\theta_{TSR}(w,u,t)$ coordinates the supply chain, then (Q_C^*,e_C^*) is the unique solution of problem (P4.1), and $E_{R,TSR}(Q_C^*,e_C^*)\geq k$ and $V_{R,TSR}(Q_C^*,e_C^*)<V_{R,TSR}(Q,e)$ for all (Q,e) that satisfy $E_{R,TSR}(Q,e)\geq k$. Then by Proposition 4.12, $(Q,e)=(Q_C^*,e_C^*)$ satisfies $E_{R,TSR}(Q_C^*,e_C^*)=k$. Hence, $(Q,e)=(Q_C^*,e_C^*)$ is also the unique solution of (P4.2).

Denote by $\tilde{\Theta}(w,u,t)$ the set of (Q,e) which satisfies $E_{R,TSR}(Q,e)>k$. $E_{R,TSR}(Q,e)$ is continuous in $Q\geq 0$ and $e\geq 0$, and $E_{R,TSR}(0,e)=-C(e)<0$. As a result, for any given $(Q_1,e_1)\in\tilde{\Theta}$, there exists $Q_2<Q_1$ such that

$$E_{R,TSR}(Q_2,e_1)=k.$$

If $(Q,e)=(Q_C^*,e_C^*)$ is the unique solution of (P4.2), then we have: $E_{R,TSR}(Q_C^*,e_C^*)=k$ and $V_{R,TSR}(Q_C^*,e_C^*)<V_{R,TSR}(Q,e)$ for all (Q,e) that satisfy $E_{R,TSR}(Q,e)=k$, and hence $V_{R,TSR}(Q_C^*,e_C^*)<V_{R,TSR}(Q_2,e_1)<V_{R,TSR}(Q_1,e_1)$. Therefore, if $(Q,e)=(Q_C^*,e_C^*)$ is the unique solution of problem (P4.2), then $(Q,e)=(Q_C^*,e_C^*)$ is also the unique solution of (P4.1) and hence $\theta_{TSR}(w,u,t)$ coordinates the supply chain.

Proposition 4.14 asserts that the supply chain coordinator (the manufacturer) can focus on a smaller problem (P4.2) (in the sense of the size of feasible solution set), instead of a bigger problem (P4.1), in coordinating the supply chain.

4.6 Concluding Remarks

We believe that the mean-risk analysis for our supply chain coordination problem with risk-averse retailer under TSR contract is meaningful. Decision makers can easily use the proposed models to determine the appropriate parameters of TSR contract for achieving coordination. Moreover, our findings are useful for decision makers to better understand the conditions under which TSR contracts can coordinate the supply chains with risk-averse retailer and sales effort-dependent demand. This will allow the decision makers to decide whether to apply, and how to apply TSR contracts in enhancing the supply chain's efficiency.

Furthermore, we believe that similar mean-risk analysis can also been studied when we consider that the supply chain has different kinds of structure. For example, we can consider a supply chain with a risk-averse manufacturer and a risk-neutral retailer, or consider a supply chain that both the manufacturer and the retailer being risk averse. We can also examine different types of supply contracts for coordinating the supply chain and consider a supply chain model with price-sensitive demand (see, for example, Chiu et al. 2011b).

References

Arcelus, F. J., Kumar, S., & Srinivasan, G. (2012). The effectiveness of manufacturer vs. retailer rebates within a newsvendor framework. *European Journal of Operational Research, 219*, 252–263.

Cachon, G. P. (2003). Supply chain coordination with contracts. In S. Graves & T. De Kok (Eds.), *Handbooks in operations research and management science: Supply chain management*. Amsterdam: North-Holland.

Cachon, G. P., & Lariviere, M. (2005). Supply chain coordination with revenue sharing contracts: Strengths and limitations. *Management Science, 51*, 30–44.

Chen, F, & Federgruen, A. (2000). *Mean-variance analysis of basic inventory models*. Working paper, Columbia University.

Chiu, C. H., Choi, T. M., & Li, X. (2011a). Supply chain coordination with risk sensitive agent under target sales rebate contracts. *Automatica, 47*, 1617–1625.

Chiu, C. H., Choi, T. M., & Tang, C. S. (2011b). Price, rebate, and returns supply contracts for coordinating supply chains with price dependent demand. *Production and Operations Management, 20*, 81–91.

Choi, T. M., Li, D., Yan, H., & Chiu, C. H. (2008). Channel coordination in supply chains with members having mean-variance objectives. *Omega, 36*, 565–576.

Elmaghraby, W., Gulcu, A., & Keskinocak, P. (2008). Designing optimal preannounced markdowns in the presence of rational customers with multiunit demands. *Manufacturing & Service Operations Management, 10*, 126–148.

Gan, X., Sethi, S. P., & Yan, H. (2004). Coordination of supply chains with risk-averse agents. *Production and Operations Management, 13*, 135–147.

Gan, X., Sethi, S. P., & Yan, H. (2005). Channel coordination with a risk-neutral supplier and a downside-risk-averse retailer. *Production and Operations Management, 14*, 80–89.

Krishnan, H., & Winter, R. A. (2011). On the role of revenue-sharing contracts in supply chains. *Operations Research Letters, 39*, 28–31.

Lian, Z., & Deshmukh, A. (2009). Analysis of supply contracts with quantity flexibility. *European Journal of Operational Research, 196*, 526–533.

Markowitz, H. M. (1959). *Portfolio Selection: Efficient Diversification of Investment*. New York: John Wiley & Sons.

Nahmias, S. (1997). *Production and operation analysis* (3rd ed.). New York: Mc Graw Hill.

Pasternack, B. A. (1985). Optimal pricing and returns policies for perishable commodities. *Marketing Science, 4*, 166–176.

Sethi, S. P., Yan, H., & Zhang, H. (2004). Quantity flexibility contract: optimal decisions with information updates. *Decision Sciences, 35*, 691–712.

Sheu, J. B. (2011). Marketing-driven channel coordination with revenue-sharing contracts under price promotion to end-customers. *European Journal of Operational Research, 214*, 246–255.

Taylor, T. A. (2002). Supply chain coordination under channel rebates with sales effort effects. *Management Science, 48*, 992–1007.

Tsay, A. A., Nahmias, S., & Agrawal, N. (1999). Modelling supply chain contracts: A review. In S. Tayur et al. (Eds.), *Quantitative models for supply chain management* (pp. 299–336). Dordrecht: Kluwer Academic Publishers.

Wu D (2011) Coordination of competing supply chains with news-vendor and buyback contract. *International Journal of Production Economics*, in press.

Yin, R., Aviv, Y., Pazgal, A., & Tang, C. S. (2009). Optimal markdown pricing: Implications of inventory display formats in the presence of strategic customers. *Management Science, 55*, 1391–1408.

Chapter 5
Mean-Risk Analysis: Conclusion, Future Research and Extensions

In the previous chapters, we discussed the use of mean-risk approach in conducting supply chain risk analysis for the single period and multiple periods supply chain inventory problems. We also explored the incorporation of mean-risk approach in the determination of the supply chain coordination contracts. From all the findings and insights generated, we have confidence to conclude that the mean-risk approach is obviously a versatile and significant way to conduct supply chain risk analysis. We believe that it has more potential than what we reviewed and discussed in the previous chapters. As a result, as a part of our conclusion, we discuss several future research directions and probable extensions for mean-risk supply chain analysis.

5.1 Expanding the Horizon

In the majority of the existing literature on stochastic supply chain analysis, the objective function is based on the expected measure (such as expected profit, and expected cost) which does not fully reflect the stochastic nature of the problem. As such, we believe that it is of paramount importance to expand the horizon and analyze the problem with respect to both the "mean" and the "risk". First, it is possible to incorporate the mean-risk measures into the optimization model and then investigate the problem (as what we examined in the previous chapters). However, this approach will inevitably complicate the problem and may lead to an analytically intractable solution in many more complex problems. This gives rise to the second approach in which one can still use the expected measure (such as expected profit) as the sole optimization objective while make use of the mean-risk approach in analyzing the supply chain performance with respect to the expected measure optimization solution. For example, Choi et al. (2003) examine the optimal two-stage inventory policy with Bayesian forecast updating and uncertain costs. They develop the dynamic optimization model and derive the optimal policy which maximizes the expected profit. After that, they employ the variance of profit as a

T.-M. Choi and C.-H. Chiu, *Risk Analysis in Stochastic Supply Chains: A Mean-Risk Approach*, International Series in Operations Research & Management Science 178,
DOI 10.1007/978-1-4614-3869-4_5, © Springer Science+Business Media New York 2012

measure to study the level of risk associated with different expected profit-to-go maximizing ordering policies. Choi et al. (2004) explore a supply chain management problem where there exists a secondary market in the form of e-marketplace. They consider the supply chain where the manufacturer offers a returns policy to the retailer. In the presence of the e-marketplace, the returned products from the primary market can be sold directly by the manufacturer to consumers via the Internet. They derive the expected profit maximizing optimal returns policy under the existence of the e-marketplace. They then further conduct a mean-variance analysis to reveal the level of risk associated with the optimal policy. Ning et al. (2011) extend Choi et al. (2004) to study the case with the markdown money policy and the variance of profit is also employed to conduct risk analysis. Notice that in Choi et al. (2003, 2004), and Ning et al. (2011), the optimization models are constructed as two-stage dynamic programs. Owing to the non-separable nature of the variance of profit, one cannot directly incorporate the variance of profit into the dynamic programme and derive the optimal policy. As such, it is practical and useful to use the variance of profit as a supplementary performance measure to reveal the level of risk associated with the expected profit maximizing optimal policy. Moreover, even for relatively simple problems, similar approach can also be taken. For instance, Choi and Chow (2008) study the quick response programme in fashion apparel with a single ordering opportunity (and hence there is no need to employ dynamic programming to solve multiple stages problem). They discuss the performance of different supply chain commitment contracts, such as price commitment policy, service-level commitment policy, and buy-back policy, which can lead to the win–win scenario between the supplier and the buyer. They propose the terminology of mean-variance win–win situation to refer to the case in which both the supplier and the buyer will be better off with the considerations of both expected profit and risk after adopting the proposed measures. With the use of variance of profit as the measure of risk, they analytically derive the conditions under which both supply chain coordination and the mean-variance win–win situation can be achieved. Recently, Chiu and Choi (2012) study the use of a rebate, returns, and wholesale pricing supply contract in coordinating supply chains with price-dependent demands. Since there exist multiple combinations of contract parameters which can coordinate the supply chain (in terms of the maximization of supply chain system's expected profit), they employ the variance of profit as a measure of risk and derive the optimal contract parameters which can minimize the level of risk of the manufacturer and maximize the expected profits of not only the manufacturer but also the whole supply chain at the same time.

As shown in the above examples, we believe that for a lot of supply chain management problems with a stochastic nature, researchers can generate many more insights by extending their works with the mean-risk consideration in two ways. The first way refers to the case when we incorporate the mean-risk objective into the respective supply chain optimization problem directly. The second way is a two-level analysis in which the first level (optimization model) still employs the expected measure (such as expected profit) as the optimization objective, while the second level then uses the risk measure (such as variance of profit) to analyze the performance of the supply chain with respect to the first level's optimization decision. We believe that the first way is especially useful for the relatively simple problems (such as the single-period models) while the second way can provide many insights for the relatively complex problems (such as the multiperiod models) and the problems with which multiple solutions for the "first level model" exist.

5.2 Information Asymmetry

In the supply chain mean-risk analysis conducted in Chap. 4, we assume that all the model parameters, as well as the risk related preference (such as the risk tolerance level, the degree of risk aversion etc.), are all publicly known to every supply chain agent. However, in a multi-echelon supply chain context, how can the upstream supplier know about the downstream retailer's risk tolerance level? As such, we have the practical information asymmetry issue in which supply chain agents would have their own private information. In the literature, supply chain management under information asymmetry has been a hot topic over the past decade. In Ha (2001), a single-supplier, single-buyer supply chain under asymmetric cost information is explored. A cutoff level policy on the buyer's marginal cost is shown to be optimal and its relationship with the model's parameters is discussed. In Corbett et al. (2004), the value of obtaining better information about the buyer's cost structure for designing supply contracts under asymmetric information is studied. They explore six cases with three increasingly general contract types and generate managerial insights regarding the value of information for different cases. Relatively recently, the supply chain coordination literature under information asymmetry has been extended in a number of ways. Some examples include: (a) studying the contracting and information sharing schemes with competition between supply chains (Ha and Tong 2008), (b) exploring a multi-period contract (called promised-lead-time contract) under information asymmetry with a focal point on comparing the profit and risk between the system with full information and the system with asymmetric information (Lutze and Özer 2008), (c) investigating a supply chain contracting problem under asymmetric production cost information by modeling the problem as a game of adverse selection (Cakanyildirim et al. 2009), and (d) proposing a menu of commitment-penalty contracts that can help provide higher certainty of demand and supply under asymmetric demand information (Gan et al. 2009). In addition, an interesting topic of whether or not the supply chain agents should share truthful information is studied in Wang et al. (2009). They find that the supply chain agents do have incentives to offer fake information in order to achieve a higher expected profit level. Notice that the above works report the analysis on information asymmetric supply chains without considering the mean-risk related issue.

Under the mean-risk framework, a couple of interesting new issues would arise. For example, consider a two-echelon supply chain in which the retailer is risk averse and the manufacturer offers a supply contract. Obviously, the retailer's degree of risk-aversion is a piece of private information which is unknown to the manufacturer. In this case, how should the manufacturer construct the mean-risk optimization model and set the parameters to achieve the optimal supply contract? More interestingly, if the retailer is willing to inform the manufacturer its own degree of risk-aversion, does the retailer have incentive to tell lies by giving a fake number on its own degree of risk-aversion (i.e., a moral hazard issue)? These are all interesting issues and in the recent literature, Wei and Choi (2010) examine parts of these in the presence of wholesale pricing and profit sharing contract. They find that the retailer does have full incentive to tell lies in many cases. As a result, in order to avoid the moral hazard issue, they suggest the manufacturer revise the contract by making it a menu of contracts in which there are multiple "wholesale price and the corresponding minimum committed quantity" pairs.

As discussed above, the information asymmetry issue is interesting and a lot of new topics and managerial insights can be explored by examining supply chain models in which information asymmetry exists among the supply chain agents. The problem naturally becomes more interesting when we conduct mean-risk analysis because the asymmetric information will potentially affect both the mean and the risk of each supply chain agent. Moreover, we will even have another source of asymmetric information, which comes from the unknown risk preference of individual supply chain agents. As a consequence, it is crystal clear that a lot of promising future research works on mean-risk supply chain analysis can be conducted under the information asymmetry setting.

5.3 More General Supply Chains

In order to show more analytical insights, the supply chains that we studied in the previous chapters are relatively simple. A natural future research direction is to investigate more general and complicated supply chains in the "longer" supply chain scenario (e.g., three or more echelons), the "wider" supply chain scenario (e.g., with multiple retailers), and the multiple periods scenario.

For the longer supply chain, it is interesting to examine how the coordination mechanism can be adjusted to achieve supply chain coordination for a three-echelon or even a longer supply chain under mean-risk models. Issues on the relationships between the performance of different channel leadership and the degrees of risk aversion of the individual supply chain agents are promising areas to study.

For the wider supply chain, it is known that coordination becomes very difficult even under the expected profit maximization case. One major challenge of this problem comes from the legal concerns on price discrimination and the presence of the Robinson-Patman Act[1] (e.g., the manufacturer cannot "discriminate" the retailers by offering different supply contracts to them separately). As a result, the use of menu of contracts becomes the "standard" treatment for this kind of problem and is proven to

[1] See Crowley (1947) for more details.

be effective (Agrawal and Seshadri 2000; Chen and Seshadri 2005). Recently, Chiu et al. (2012a) examine the supply chain coordination problem with a risk neutral manufacturer (called a global brand) which supplies to multiple risk-averse retailers. They consider the case when these retailers differ by having different degrees of risk aversion (formulated under the mean-risk model). They analytically prove that an innovative menu of contracts, which include the addition of various contract parameters termed as the risk-level indicator and the separation indicator, can help coordinate the supply chain. Based on the results of Chiu et al. (2012a), we hypothesize that in order to coordinate a wider supply chain with risk averse agents (under the mean-risk model), we can employ the menu of contracts approach and include new contract parameters to enhance its versatility. Nevertheless, to provide solid evidence on this hypothesis, more future research in this area is needed.

For the multi-period supply chains, risk analysis becomes more complicated and challenging. In Chap. 3, we discussed the case with the (R, nQ) policy while the use of variance measure is only confined to each individual period, but not multiple periods together. The same situation under mean-risk framework in the cost domain is explored by Chen and Federgruen (2000). Recently, there are some attempts, including Chen et al. (2007), and Choi et al. (2011), which study the multi-period inventory control problems with more considerations. To be specific, Chen et al. (2007) propose a framework based on expected utility (exponential) approach to examine inventory control problems with risk averse decision makers in a multi-period setting. They find that the optimal policy's structures under the risk averse case and the risk-neutral case are very similar. Their computational results further demonstrate that the optimal policy is rather insensitive to small changes in the level of risk aversion of the decision maker. Choi et al. (2011) investigate a finite multiple-period (N-period) inventory control problem under a mean-variance framework. They construct a mean-variance optimization objective for a periodic review base-stock inventory system. Owing to the non-separable nature of the variance of (N-period total) profit measure in the objective function, dynamic programming cannot be directly applied to solve the problem. As a result, Choi et al. (2011) propose a new approach to solve this challenging problem by creating an auxiliary problem and finding the conditions under which its solution and the solution of the original mean-variance optimization problem will converge.

Inspired by the above work, future research can be conducted in two directions. Direction one focuses on the advancement of both the modeling and the optimization techniques to deal with multi-period mean-risk inventory control problems. Many recent advances in mean-variance portfolio optimization in finance (e.g., Cui et al. 2012; Li and Ng 2000) can be good references. Direction two refers to the analysis of multi-echelon supply chain management problem in the setting with multiple periods and mean-risk analysis. Issues such as channel coordination with risk averse agents in a multi-period setting and the use of dynamic (e.g., state dependent) supply chain contracting mechanism[2] can be explored.

[2] An example of a dynamic supply contract can be found in Chow et al. (2012).

5.4 Behavioral Research

The traditional supply chain management literature with mathematical modeling research focuses on generating insights and developing solution schemes via standard analytical studies such as employing optimization methods and game-theoretic analysis. Under the mean-risk framework, it basically means constructing, analyzing, and solving mathematical models with both mean and risk considerations. In recent years, there is an observed trend that makes use of experiment-based behavioral research to supplement and strengthen analytical studies. For example, in Su (2008), the newsvendor problem with a boundedly rational decision maker is studied. Su's decision model under bounded rationality is based on the quantal choice model, which considers the case where the decision is not always the "best" but usually a "better" one. Empirical evidence of bounded rationality is provided by using a data set of newsvendor-type decisions made by individual human subjects. Systematic biases are identified and the impacts of these biases on supply chain problems such as contracting, bullwhip effect, and risk pooling are all proposed by Su (2008). In the supply chain channel coordination literature, the use of behavioral experiments as a research method is also popular. For example, in Chen et al. (2008), a dual sales channel (online sales and bricks-and-mortar store) management problem with service competition between them is studied. After conducting analytical studies and developing the optimal duel channel strategies, controlled experiments with human subjects are conducted to investigate whether the model makes reasonable predictions of human behaviors. Recently, Davis (2012) studies a supply chain "pull" contracting mechanism in which an upstream supplier incurs the holding cost of inventory and uses it to directly satisfy a downstream retailer's demand. He investigates three such pull contracts, namely the pure wholesale pricing contract and two other coordinating contracts, in a controlled laboratory setting. Their empirical results indicate that counter-intuitively, the theoretically proven "beneficial" coordinating contracts need not outperform the wholesale pricing contract. They explain this behavioral finding by exploring a risk aversion model and a model of anticipated regret. More related works on supply chain management related behavioral research can be found in Wu and Chen (2012), Kalkanci et al. (2011), Gino and Pisano (2008) and the references therein.

Undoubtedly, many influential and interesting new findings can be revealed via behavioral research. As a consequence, it is natural to conduct future research with human subjects based behavioral experiments to verify, refine, and extend all kinds of analytically constructed mean-risk supply chain models. For example, whether the model assumption on perfectly rational "risk averse" decision making mechanism under our mean-risk framework is valid or not[3] in the real world is an important topic which deserves full attention in future research.

---

[3] As a comparison, decision makers may only be "boundedly rational" (see Su (2008) and Wu and Chen (2012) for more discussions).

Table 5.1 A summary of the proposed future research directions and related references

	Details	Related references
Section 5.1	From expected profit/expected cost model to mean-risk model with two approaches (as follows).	
	1. Incorporates the mean-risk objective into the respective supply chain optimization problem.	Vaagen and Wallace (2008), Wu et al. (2010), Liu et al. (2011), Chiu et al. (2012a)
	2. A two-level analysis in which the first level (optimization model) still employs the expected measure as the optimization objective; the second level uses the risk measure to analyze the performance of the supply chain with respect to the first level's optimization decision.	Choi and Chow (2008), Ning et al. (2011), Chiu and Choi (2012)
Section 5.2	1. Unknown risk preference.	Choi et al. (2008a)
	2. Moral hazard issue.	Wang et al. (2009), Wei and Choi (2010)
	3. Asymmetric information on demand distribution, cost-revenue parameters in the supply chain between supply chain agents.	Ha (2001), Corbett et al. (2004), Ha and Tong (2008), Lutze and Özer (2008), Cakanyildirim et al. (2009), Gan et al. (2009)
Section 5.3	1. Longer supply chains: Relationships between performance of channel leadership and risk aversion.	Chiu et al (2012b)
	2. Wider supply chains: Coordination by innovative menu of contracts and dynamic contracts.	Agrawal and Seshadri (2000), Chen and Seshadri (2005), Chiu et al. (2012a), Chow et al. (2012)
	3. Mean-risk analysis of multi-period supply chains: (a) New optimization methods. (b) Coordination mechanism.	Chen and Federgruen (2000), Choi et al. (2011)
Section 5.4	Verify, refine and extend mean-risk analytical supply chain models and coordination problems by human-subject behavioral experiments.	Su (2008), Davis (2011), Kalkanci et al. (2011), Wu and Chen (2012)

Table 5.1 summarizes the major future research directions and the closely related references as discussed above.

References

Agrawal, V., & Seshadri, S. (2000). Risk intermediation in supply chains. *IIE Transactions, 32*, 819–831.

Cakanyildirim, M., Gan, X., & Sethi, S. (2009). *Contracting and coordination under asymmetric production cost information*. Working paper, University of Texas at Dallas.

Chen, F., & Federgruen, A. (2000). *Mean-variance analysis of basic inventory models*. Working paper, Columbia University.

Chen, K. Y., Kaya, M., & Özer, O. (2008). Dual sales channel management with service competition. *Manufacturing and Service Operations Management, 10*, 654–675.

Chen, Y. J., & Seshadri, S. (2005). Supply chain structure and demand risk. *Automatica, 42*, 1291–1299.

Chen, X., Sim, M., Simchi-Levi, S., & Sun, P. (2007). Risk aversion in inventory management. *Operations Research, 55*, 828–842.

Chiu, C. H., & Choi, T. M. (2012). *Optimal price, rebate, and returns supply contracts for channel coordination with price-dependent demands and risk consideration*. Working paper, The Hong Kong Polytechnic University.

Chiu, C. H., Choi, T. M., Hao, G., & Li, X. (2012a). *Coordinating supply chains with a global-brand and multiple risk-averse retailers*. Working paper, The Hong Kong Polytechnic University.

Chiu, C. H., Choi, T. M., Li, X., & Yiu, C. K. F. (2012b). *Supply chain coordination under different channel leaderships and information asymmetry with stochastic price-dependent-demands*. Working paper, The Hong Kong Polytechnic University.

Choi, T. M. (2011). Coordination and risk analysis of VMI supply chains with RFID technology. *IEEE Transactions on Industrial Informatics, 7*, 497–504.

Choi, T. M., Chiu, C. H., & Fu, P. (2011). Periodic review multi-period inventory control under a mean-variance optimization objective. *IEEE Transactions on Systems, Man, and Cybernetics, Part A - Systems and Humans, 41*, 678–682.

Choi, T. M., & Chow, P. S. (2008). Mean-variance analysis of quick response program. *International Journal of Production Economics, 114*, 456–475.

Choi, T. M., Li, D., & Yan, H. (2003). Optimal two-stage ordering policy with Bayesian information updating. *Journal of the Operational Research Society, 54*, 846–859.

Choi, T. M., Li, D., & Yan, H. (2004). Optimal returns policy for supply chain with e-marketplace. *International Journal of Production Economics, 88*, 205–227.

Choi, T. M., Li, D., & Yan, H. (2008a). Mean-variance analysis of a single supplier and retailer supply chain under a returns policy. *European Journal of Operational Research, 184*, 356–376.

Chow, P. S, Choi, T. M., & Cheng, T. C. E. (2012). Impacts of minimum order quantity constraint on a quick response supply chain. *IEEE Transactions on Systems, Man, and Cybernetics, Part A - Systems and Humans* (doi:10.1109/TSMCA.2012.2183351), in press.

Corbett, C. J., Zhou, D., & Tang, C. S. (2004). Designing supply contracts: Contract type and information asymmetry. *Management Science, 50*, 550–559.

Crowley, P. J. B. (1947). Equal price treatment under the Robinson-Patman Act. *University of Pennsylvania Law Review, 95*, 306–343.

Cui, X., Gao, J., Li, X., & Li, D. (2012). *Optimal multiperiod mean-variance policy under no-shorting constraint*. Working paper, The Chinese University of Hong Kong.

Davis, A. M. (2012). An experimental investigation of pull contracts. *Manufacturing and Service Operations Management*, forthcoming.

Gan, X., Sethi, S. P., & Zhou, J. (2009). *Commitment-penalty contracts in drop-shipping supply chains with asymmetric demand*. Working paper, The Hong Kong Polytechnic University.

Gino, F., & Pisano, G. (2008). Toward a theory of behavioral operations. *Manufacturing and Service Operations Management, 10*, 676–691.

Ha, A. Y. (2001). Supplier-buyer contracting: Asymmetric cost information and cutoff level policy for buyer participation. *Naval Research Logistics, 48*, 41–64.

Ha, A. Y., & Tong, S. (2008). Contracting and information sharing under supply chain competition. *Management Science, 54*, 701–715.

Kalkanci, B., Chen, K. Y., & Erhun, F. (2011). Contract complexity and performance under asymmetric demand information: An experimental evaluation. *Management Science, 57*, 689–704.

Li, D., & Ng, W. L. (2000). Optimal dynamic portfolio selection: Multiperiod mean-variance formulation. *Mathematical Finance, 10*, 387–406.

Liu, N., Choi, T. M., Yuen, M. C. W., & Ng, F. (2011). Optimal pricing, modularity and return policy under mass customization. *IEEE Transactions on Systems, Man, and Cybernetics, Part A*, (doi: 10.1109/TSMCA.2011.2170063), published online.

Lutze, H., & Özer, Ö. (2008). Promised lead-time contracts under asymmetric information. *Operations Research, 56*, 898–915.

Ning, Z., Choi, T. M., Xie, C., Xie, L., & Dai, J. (2011). Impact of e-marketplace on supply chain under the markdown policy. *Supply Chain Management: An International Journal, 16*, 409–418.

Su, X. (2008). Bounded rationality in newsvendor models. *Manufacturing and Service Operations Management, 10*, 566–589.

Vaagen, H., & Wallace, S. W. (2008). Product variety arising from hedging in the fashion supply chains. *International Journal Production Economics, 114*, 431–455.

Wang, J. C., Lau, H. S., & Lau, A. H. L. (2009). When should a manufacturer share truthful manufacturing cost information with a dominant retailer? *European Journal of Operational Research, 197*, 266–286.

Wei, Y., & Choi, T. M. (2010). Mean-variance analysis of supply chains under wholesale pricing and profit sharing scheme. *European Journal of Operational Research, 204*, 255–262.

Wu, D. Y., & Chen, K. Y. (2012). *Supply chain contract design: Impact of bounded rationality and individual heterogeneity*. Working paper, University of Kansas.

Wu, J., Wang, S., Chao, X., Ng, C. T., & Cheng, T. C. E. (2010). Impact of risk aversion on optimal decisions in supply contracts. *International Journal of Production Economics, 128*, 569–576.

Index

T.-M. Choi and C.-H. Chiu, *Risk Analysis in Stochastic Supply Chains: A Mean-Risk Approach*, International Series in Operations Research & Management Science 178,
DOI 10.1007/978-1-4614-3869-4, © Springer Science+Business Media New York 2012

MIX
Papier aus verantwortungsvollen Quellen
Paper from responsible sources
FSC® C105338

If you have any concerns about our products,
you can contact us on
ProductSafety@springernature.com

In case Publisher is established outside the EU,
the EU authorized representative is:
Springer Nature Customer Service Center GmbH
Europaplatz 3, 69115 Heidelberg, Germany

Printed by Libri Plureos GmbH
in Hamburg, Germany